Slugs, Pests and Diseases

Slugs,
Pests
and
Diseases

READER'S
DIGEST

CONTENTS

The Healthy Garden

A HEALTHY GARDEN ALL YEAR ROUND

Directory of Pests and Diseases

Special features

the healthy garden

8 The healthy garden

THERE IS NOTHING YOU CAN DO TO STOP PESTS AND DISEASES CROSSING YOUR BOUNDARY. INSTEAD, ACCEPT A LOW LEVEL OF INFECTION, AND TAKE A FEW SIMPLE PRECAUTIONS TO PREVENT PROBLEMS GETTING OUT OF HAND.

Vigorous, well-grown plants that are given plenty of space to develop naturally are more likely to withstand disorders than overcrowded, starved and dry plants, so encourage the good health of your plants by satisfying their needs: ■ Water them before they are too dry ■ Feed adequately to sustain steady growth ■ Reduce competition from weeds ■ Keep the garden tidy: clear away dead plants, check soil acidity and drainage, and avoid over-feeding.

Learn to identify the most common garden pests and to distinguish them from gardeners' allies such as ladybirds, lacewings, centipedes, hoverflies and ground beetles and their larval, immature forms. And remember that some pests, such as froghoppers ('cuckoo spit'), are simply a nuisance and may be ignored, whereas others like the beautiful lily beetle can be a serious threat.

Butterflies and other garden friends are attracted by flat-headed blooms, such as sedums, and when resident in your garden, will act as your very own, natural pest-control army, so encourage them by growing a wide range of the right plants and limiting your use of chemical treatments.

GLORIOUS DISPLAYS AT THE GARDEN CENTRE MAKE IT ALL TOO TEMPTING TO FILL YOUR TROLLEY WITH IMPULSE BUYS, BUT SPEND TIME CHOOSING SUITABLE, HEALTHY SPECIMENS AND YOU WILL SAVE TIME IN THE LONG RUN. POOR AND NEGLECTED PLANTS NEED MORE AFTERCARE AND CAN BRING PROBLEMS, SUCH AS WEEDS, PESTS AND DISEASES.

A good-quality plant will have been raised under ideal conditions and delivered recently to the shop. It should be established in its pot, but not pot-bound – the roots should hold the compost but not be a solid mass, tightly coiled around the bottom of the pot. Don't be shy about carefully turning plants out of their pots to inspect the roots.

Selecting plants

Plants left on garden centre display benches for too long can run short of water and nutrients in their small pots. This can lead to wilting, premature flowering and discoloured foliage. Overwatering can also be a problem – water-logged roots cannot survive for long and are vulnerable to fungal disease. Liverwort or moss on the surface of the compost is a sign that it has been kept too wet.

Health check

Prevent problems in your garden by inspecting shoot tips, young leaves (including the undersides) and roots for signs of pests and diseases. Check fuchsias, alcea (hollyhocks) and roses for rust spots beneath the leaves, alpines and herbaceous plants for grey mould, and look for the tell-tale 'webs' of spider mite under the leaves of fruit trees and plants sold under cover. Also be alert for limpet-like scale insects on woody plants and powdery mildew on fruit trees, roses and perennials.

Vine weevil is a serious problem on many types of garden plant and is difficult to treat.

What to look for

■ **HERBACEOUS PERENNIALS** Buy small (9cm) pots in spring, but larger pots in summer, which have more space for the roots to have developed. Hardy perennials can be bought in autumn even if top growth has started to die back, but check roots first for signs of damage.

■ **CLIMBERS** Check stems near the base of the plant because they are easily damaged.

■ **TREES & SHRUBS** Look for a good shape that is not lop-sided or leggy to save having to prune it into shape. Check the rootball to make sure that the plant is not badly pot-bound.

■ **CONIFERS** Buy only named varieties to avoid planting trees that will grow too fast and too large for your site. True dwarf varieties cost more, but they are essential in a restricted space.

■ **PATIO & BEDDING PLANTS** Don't buy too early in the year, when frosts are still a risk. Bedding plants are sold at various stages of maturity. The more mature the plant, the less nurturing it will need, but you will get less choice and it will cost more.

■ **BULBS** For both spring and summer-flowering bulbs, those bought as plants about to flower give instant results and are often more successful than dry bulbs.

Check for a healthy rootball before you buy

The adult makes notches on leaf edges, and if this is on young foliage then they have probably laid eggs in the compost too and the resulting grubs will eat the roots. Also avoid plants with unnatural mottling or streaking of foliage because these are symptoms of virus disease for which there is no cure.

Pull off any weeds and remove the top few centimetres of compost before planting the new plant in your garden.

10 Preventing pests and diseases

Taking action to prevent pests and diseases is more time-effective than controlling them once they have taken hold. Good garden hygiene will go a long way to avoiding most problems. So will growing a wide diversity of plants, because too much of one type of plant can provide a haven for its predators.

Bring only healthy stock into your garden, purchased from tidy, weed-free nurseries and garden centres, and inspect plants and bulbs before buying them (see page 9).

Prevent carrying over diseases by washing pots and seed trays with a garden disinfectant. Rinse them thoroughly before use.

Garden hygiene
Remove the remains of old, rotting or infected plants promptly. Dig up and destroy any plants with virus symptoms so that the problem does not spread to other plants. Keep on top of weed control, particularly near the vegetable plot, because weeds often act as a host for pests.

Once plant tissues start to rot, secondary problems with fungi and bacteria can take hold. Careful planting and watering can prevent rotting. For example, certain large seeds and some tubers are best planted on their side. Some alpines can be protected

from rot with a collar of fine gravel, which allows water to drain away quickly.

Plants growing too close together compete for light, water and nutrients, and cannot grow to their potential. Overcrowding also prevents good air circulation, allowing fungal diseases to take hold. Overcrowded plants are difficult to inspect and care for, so problems can become serious before they are noticed.

Cuts on plant tissue can provide an entry point for pests and diseases. Take care not to damage plants when hoeing, mowing or using a nylon-line trimmer. Use sharp tools when pruning and make clean cuts just above a bud to prevent die-back. Make sure that ties on trees and climbers are secure but do not cut into the plants.

In the warm conditions inside a greenhouse, young plants are particularly at risk of attack by pests and diseases. Raise seeds, young plants and cuttings in clean pots. Use fresh compost and tap water rather than water collected in a water butt because many problems are spread by infected water. While plants growing in the ground

Anti-insect compost
SPECIAL POTTING COMPOST containing a systemic insecticide not only supplies plants with a growing medium and nutrients, but also protects them from pests for up to a year. Once planted, the plant absorbs the insecticide through its roots and transports it to all its parts. If an insect takes a bite or sucks the sap it will die. The compost protects plants from caterpillars and vine weevil grubs, and all sap-sucking insects, such as aphids, whitefly and spider mite. It is widely available.

or in containers can be watered from stored water in butts, it is important that the butt be kept covered and that it is cleaned out once a year.

Honey fungus
If honey fungus is present in the soil, it can be a serious and time-consuming problem. Golden brown toadstools at the base of an old tree may indicate an infection, but to make sure, take a sharp knife and lift up an area of bark near the base of the tree. Look for white, paper-like growths or black growths that look like bootlaces, either under the bark or in the surrounding soil. Early leaf-fall from deciduous plants and die-back of branches can also be signs of fungi. If you suspect a honey-fungus infection in your garden, seek specialist advice.

Several garden plants are particularly susceptible to honey fungus, including birch, buddleja, cotoneaster, flowering currant, hop, Lawson cypress, Leyland cypress, lilac, maple, privet, rhododendron, rose, viburnum, willow, most fruit trees and soft fruit bushes. Avoid planting these if there has been a honey-fungus infection in your garden. Plants that show some resistance to the fungus include bamboo, beech, cherry, laurel, holly, juniper, larch and yew.

Weather watch
IN SPRING If it has been a mild winter, watch out for aphid attacks in spring. If you spot them early, rub out colonies between finger and thumb or apply a ready-to-use spray insecticide.
IN SUMMER When the weather is warm, pest populations can soar in a matter of days. Humid days also create the perfect environment for diseases to take hold in overcrowded beds.
IN AUTUMN AND WINTER Clear up fallen leaves if they start to show symptoms of disease or are stifling other plants or the lawn.

First-aid kit for your plants
YOU DON'T NEED a shed full of chemicals to control most common pests and diseases, but it is worth having an arsenal of ready-to-use sprays for tackling problems early. Keep a magnifying glass, notepad and pencil handy to jot down the names of pests, the treatment given and whether it worked, so that you can learn as you go along.

Insecticides
Choose a general insecticide based on permethrin or thiocloprid. Systemic treatments are transported throughout the plant and do not have to be sprayed directly onto the pest. Organic gardeners should look for products based on pyrethrum instead.

Slug pellets
If your plants suffer from slug damage every year, prevent attacks by applying slug pellets around susceptible plants before the new leaves emerge in spring.

Fungicides If you grow roses, choose a chemical 'cocktail' that will control rust, mildew and blackspot. If you grow edible crops, check that the product is suitable. Organic gardeners can use fungicides based on copper or sulphur, but they are not so effective.

Aphicides Aphids are so troublesome that it is worth having a ready-to-use insecticide to control outbreaks when you see them.

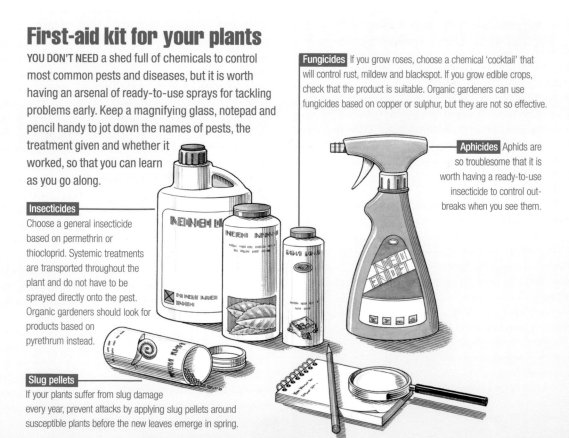

12 Beneficial creatures

IN THE WILD, ALL PESTS HAVE NATURAL ENEMIES THAT KEEP THEM IN CHECK. ENCOURAGE THESE CREATURES TO TAKE UP RESIDENCE IN YOUR GARDEN AND THEY WILL FIGHT YOUR PESTS FOR YOU BY PROVIDING THE RIGHT SORT OF HABITAT AND AVOIDING THE USE OF CHEMICALS THAT CAN CAUSE THEM HARM.

By handing over the control of garden pests to their natural predators you will rarely need to use sprays. But you must encourage the right mix of wildlife in your garden by providing conditions to suit them. A low level of pests has to be tolerated to provide a food source for the predators.

Beating aphids
Ladybirds are commonly known as the 'gardener's friend' because of their appetite for aphids, but many insects or their larvae – lacewings, parasitic wasps and hoverflies, for example – also eat vast quantities. Any nectar-producing and pollen-rich plants are useful for attracting predators; *Convolvulus tricolor* and the poached-egg plant (*Limnanthes douglasii*), for example, are good for attracting hoverflies, which eat pollen before laying their eggs among aphid colonies.

Ground beetles feed at night on aphids, the eggs of carrot fly and cabbage root fly, and slugs. Encourage them by providing ground cover to hide in during the day. Use only short persistence products, like derris, and do not use slug pellets containing methiocarb, which is toxic to ground beetles.

Encourage natural predators such as hoverflies, whose larvae feed on aphids.

Combating slugs
Frogs and toads eat slugs and insects – some of which are beneficial. To encourage them, build a pond with gently sloping sides surrounded by vegetation. Hedgehogs will also eat slugs plus caterpillars. They are difficult to encourage into a garden, but a logpile will provide a suitable place for winter hibernation or you can buy a special hedgehog house. Before lighting a bonfire, it is particularly important to look for sheltering hedgehogs.

Busy earthworms
Perhaps the most beneficial creatures to the busy gardener are earthworms, which help to improve the soil by incorporating organic matter and opening it up with their tunnels. Chemicals were once used to clear lawns of wormcasts, but modern gardeners should just sweep off casts with a broom before mowing. Brandling worms are found in compost heaps, where they break down organic material.

New Zealand flatworms, which eat earthworms, have been found in some parts of the UK and could threaten soil fertility if they become established. They are around 15cm long and very flat, and are found under stones. Squash any you find.

Attracting birds
Many garden birds eat pests. For example, blue tits eat caterpillars and aphids; starlings prey on grubs; sparrows, great tits and wrens consume insects; while song thrushes eat snails and slugs. They are attracted to gardens on the edges of woodlands and where there is water. If you have a cat, put a bell around its neck so the birds can hear it coming.

The best way to attract birds into the garden is to supply water, feed them in winter and put up nestboxes. You can also encourage them by growing plants with berries and seedheads.

Plants that bear berries include berberis, cotoneaster, elderberry, honeysuckle, ivy, pyracantha, *Skimmia japonica*, callicarpa, pernettya and viburnum. Seed eaters, including greenfinches and goldfinches, will be attracted by seed-bearing plants, such as ornamental grasses and sunflowers, while insect-eating birds, such as sparrows and great tits, will come to your garden if you grow nectar producing plants that attract pollinating insects. Plant alyssum, buddleja, honesty, lavender, nicotiana, sedum, sweet rocket and thyme.

Feed birds in winter with a commercial wild-bird food or put out scraps of breakfast cereals, flaked oats, dried fruit, bacon rind, cheese, stale cake or cooked potato. But do not put out salty food, dry bread, uncooked rice or desiccated coconut.

Beneficial creatures

Ground Beetles

Scurrying, big black beetles that are more often seen at night. Both adults and their larvae are useful predators of slugs, caterpillars and aphids.

Hoverflies

A group of true flies that look like bees and wasps. Their greeny brown larvae, 12mm long, feed on aphids and can eat up to 100 a day.

Hedgehogs

They eat caterpillars, beetles and slugs at night. If they visit your garden, put out tinned dog food rather than milk. A log pile might provide suitable shelter and a place to hibernate.

Rove beetles

This group includes the distinctive looking devil's coach horse beetle (above). Both adults and their larvae are active predators of soil grubs, insects and slugs.

Ladybirds

A single larva can eat 500 aphids so they are worth encouraging. Two-spot and seven-spot ladybirds are the most common.

Wasps

Many solitary wasps are beneficial against aphids and caterpillars. Provide egg-laying sites by drilling 5-10mm holes in posts.

Centipedes

These fast-moving creatures eat all types of small insects. Do not confuse them with the slower millipedes, which feed on plant roots. Centipedes have one pair of legs per body segment.

Earthworms

Worms pull organic matter into the soil, which saves you having to dig it in. They also open up the structure of the soil, aiding drainage and making it easier for plant roots to establish.

Bats

Night feeders of nocturnal flying insects. Encourage them with bat boxes where they can roost. They are under threat, mainly due to poisons in the chemical treatment of roof timbers.

Frogs, toads and newts

These amphibians prey on slugs, flies and other insects. They need a pond in which to breed and will return to it year after year.

Lacewings

Adults are bright green and have large, see-through wings and long antennae. Adults and their larvae feed on aphids. You can buy chambers to protect them and lacewing eggs.

Slow-worms

These legless lizards up to 30cm long eat the small greyish slugs that feed on the soft growth of young plants and vegetables.

14 Organic treatments

IT IS TEMPTING TO REACH STRAIGHT FOR THE SPRAY GUN AND THE CHEMICALS WHEN YOU FIRST NOTICE SIGNS OF PROBLEMS IN YOUR PLANTS, BUT THERE ARE MANY ORGANIC TREATMENTS THAT OFFER AN EQUALLY EFFECTIVE SOLUTION THAT IS ALSO LESS DAMAGING TO THE OVERALL BALANCE OF NATURE IN YOUR GARDEN.

Choosing plant varieties that are resistant to pests and diseases and that enjoy the conditions in your garden will go a long way towards removing the need for chemical intervention. Good growing methods – sow or plant at the right time and tend plants regularly so that they never lack for food, water and growing room – will also help, as will growing a wide diversity of plants and using physical barriers, such as fleece, to keep pests at bay.

It is also vital that your soil is in the best possible condition – plants forced to struggle for survival in poor, undernourished soils will be vulnerable to attack.

Healthy soil and plants

To improve soil fertility and structure, apply well-rotted compost and manure each year and protect the soil from the elements by growing green manure on empty plots.

If you are starting with poor, badly fed soil, you may need to add extra nutrients in the form of organic fertilisers until it improves. The table on the right will help you to boost a particular element, or you could choose a compound organic fertiliser, like blood, fish and bone meal, for a general pick-me-up for the garden. Only use these fertilisers when absolutely necessary and only as supplements to, not substitutes for, manure and compost.

Soil tonics

A testing kit will reveal any deficiencies in your soil's mineral levels. Follow this guide to restore the balance naturally.

Element	Source
CALCIUM (Ca)	Dolomite limestone
	Gypsum
NITROGEN (N)	Dried blood
	Fish meal
	Hoof and horn meal
PHOSPHATE (P)	Bone meal
	Rock phosphate
POTASSIUM (K)	Rock potash
	Wood ash
TRACE ELEMENTS	Calcified seaweed
	Dried animal manures
	Liquid animal manures
	Liquid seaweed
	Seaweed meal

Safer sprays for the organic gardener

Most chemicals should be avoided in the organic garden, but there are some sprays that are approved for use – even though they may also kill beneficial insects.

Ingredient	Origin	Problems controlled
COPPER	Naturally occurring element	Diseases including potato blight and damping off
PYRETHRUM	Made from the flowers of *Tanacetum cinerariifolium*	Wide range of pests including aphids and caterpillars
QUASSIA	Derived from the bark of a tree, *Picrasma quassioides*	Many leaf pests, especially aphids
ROTETONE (DERRIS)	Made from powdered roots of various tropical plants	Wide range of pests, including aphids, caterpillars, sawfly and thrips
SOAPS	Made from organic fatty acids	Wide range of pests, including aphids, red spider mite and whitefly
SULPHUR	Naturally occurring element	Fungus diseases (May damage some plants)

fertiliser solution (left), that can be used for combating pests, treating infection and boosting the health of your plants.

Most importantly, learn to distinguish between friends and foes. Never assume that an insect or other creature that appears in your garden for the first time is the scout of an invading army. It is much more likely to be a friend that has come to protect you from invasion and to enjoy the safety of your pesticide-free garden.

Making liquid fertiliser

1 Shear off the fresh young growth of comfrey or nettles in spring.
2 Pack the leaves into a net bag and tie it at the top.
3 Suspend this in a tank or bucket of water for 10–14 days, squeezing the bag occasionally. Dilute the resulting liquor to look like weak tea before applying.

Going green

To be successful, you must adopt the organic system throughout the garden. Banning chemicals from only the kitchen garden may provide you with vegetables that are free of pesticide residues but it will do nothing to restore the balance of nature that is essential in the control of pests and diseases. If you continue to spray insecticide on your roses, for example, you risk killing not only greenflies, but also the beneficial insects that were on their way to pollinate your vegetables.

Initially, while predators and parasites are assembling their troops, you may need to help them to keep pests under control. Do not be tempted to resort to chemical pesticides that will kill friends and foes alike, but choose a spray from the table opposite, all of which are suitable for use in an organic garden. There are also many homemade treatments, such as the simple

Homemade cures for common problems

BEATING APHIDS WITH GARLIC A garlic infusion is an effective treatment against many common garden pests. Roughly chop 25g of garlic cloves and place them in a heat-proof bowl. Bring 1 litre of water to the boil then pour it over the garlic. Cover the bowl and leave the mixture to infuse for 15 minutes. Strain the liquid through a large sieve. Spray the garlic infusion, undiluted, on plants susceptible to aphids every two to three weeks.

NETTLE MASH FOR COMMON PESTS Stinging nettles also make an excellent spray for killing insect pests. Chop 200g of fresh stinging nettle leaves and place them in a large bowl with 2 litres of cold water. Keep the leaves submerged by placing a weighted plate on top and leave them to infuse for 24 hours. Strain the liquid through a large sieve, dilute the extract (1:5) and spray it on your plants as soon as you notice a pest.

TREATING FUNGUS Fungal infections will spread quickly in warm, wet weather, so treat the problem as soon as you notice it with this simple horsetail decoction. Chop 500g of fresh horsetail 'needles' and stems and place in a large bucket. Pour in 5 litres of cold water and leave the mixture to infuse for 24 hours. Then transfer it to an old cooking pan and simmer over a low heat for 15–30 minutes. Leave the brew to cool then strain through a fine sieve, pressing the herbs through with the back of a wooden spoon. Dilute this decoction with water (1:5) and use it to spray plants infected with fungus on three consecutive days.

16 | Using chemicals safely

GARDEN CHEMICALS OF ALL KINDS SHOULD BE TREATED WITH CARE AND RESPECT. USE THEM ONLY AS A LAST RESORT AND ALWAYS FOLLOW THE MANUFACTURER'S INSTRUCTIONS TO THE LETTER. MOST TREATMENTS ARE DESIGNED FOR SPECIFIC PROBLEMS AND PLANTS AND SHOULD ONLY BE USED FOR THE PURPOSE FOR WHICH THEY WERE INTENDED.

Where possible, choose non-persistent soap-based insecticides. Systemic insecticides, which are absorbed by the plant, treat sap-sucking insects, such as aphids, as well as difficult problems, such as scale, mealy bugs and woolly aphids. Contact insecticides coat the surface over which the insects move or are applied directly to the pest and are effective against chewing pests, such as caterpillars.

Insecticides are available as liquids, dusts, powders or baits. Sprays (the most effective against sap-suckers) are expensive but the easiest to apply. Dusts and baits are best for controlling ants and other crawling insects.

Using fungicides
Your plants will be far less susceptible to fungal diseases if you choose varieties with some built-in resistance, give plants the light, space and food they require and are scrupulous about garden hygiene. If your plants do suffer from fungal attack, use fungicides sparingly – an overdose can be fatal. Spray from the bottom of the stem upwards, making sure that you coat the undersides of the leaves, where most diseases occur.

Some plants are sensitive to the chemicals used in fungicides and their leaves may drop or develop spots and scorches. Check the product label before you start for warnings of plants that are known to be susceptible. If in doubt, test a small area of the plant or a single plant in a group first.

Most fungicides do not actually kill fungi, but prevent the spores from spreading. At the first signs of a fungal disease, prompt spraying of the whole plant and any others of the same type standing nearby should contain the problem.

Safe storage
Most garden chemicals have a relatively short shelf life. From time to time check all bottles and packages and dispose of those you no longer require by contacting the local council environmental health department or your local waste and recycling centre. Never pour chemicals down the drain. Do not use empty containers to store other things and never store chemicals in containers other than those in which they were supplied. Keep them well away from children and animals, on a high shelf in a locked shed or garage.

Using insecticides

MOST INSECTICIDES kill all insects, helpful and harmful alike. To minimise risk to beneficial creatures, such as ladybirds, spray in the early morning or late evening when fewer of them are active in the garden. Restrict the treatment to plants where a pest or disease is visible.
- **Use a small sprayer** and mix no more than is required to do the job, or buy ready-to-use products.
- **Watch the weather** and spray only when there is little or no wind and the weather is fine.
- **Spray evenly** over the plants you are treating, particularly on the undersides of leaves. Since eggs are rarely destroyed you may need to repeat the treatment to catch the next generation of pests.
- **Wear gloves**, long sleeves, trousers and stout shoes as you work.
- **Do not drink** eat or smoke and thoroughly wash your hands and face once you have finished.
- **Keep children and pets** away from the spraying area while you are working.
- **Spray leftover insecticide** onto an area of bare ground or a gravel path. Do not leave it in the sprayer for another time.

A healthy garden all year round 17

EVERY SEASON BRINGS ITS OWN POTENTIAL PROBLEMS WITH PESTS AND DISEASES. FOLLOW THIS SEASONAL ADVICE TO AVOID THEM AS FAR AS POSSIBLE AND TO HELP YOU TO MAINTAIN A HEALTHY AND FLOURISHING GARDEN ALL YEAR ROUND.

Early spring

Jobs for early spring

Start the year with good intentions and conduct a thorough inspection of your garden once the worst of the winter weather has passed. There are lots of ways in which you can help to avoid pests and diseases taking hold in your garden.

■ **Tidy beds and borders,** clearing away plant debris and cutting dead stems back to ground level. Divide large clumps of perennials to prevent overcrowding – pests and diseases spread fast in cramped conditions.

■ **Check all supports and plant ties** and loosen or replace any that are tight.

■ **Continue to protect against frost** for plants of borderline hardiness. Remove the covers in mild weather so that plants can start to harden off.

■ **Watch for early signs** of pests and diseases, especially in heated greenhouses.

■ **Check stored bulbs and tubers** to make sure that they are still in good condition, and start them into growth at the appropriate time so that you do not put them under undue stress.

Aphids cluster on fresh new shoots and buds in spring.

Positive health care Early
spring is an ideal time to check your plants over for signs of ailments and other potential problems. Frost and wind damage should be obvious by winter's end, allowing you to do any essential remedial pruning, such as cutting out dead and diseased wood or disfigured evergreen foliage. The first signs of spring and summer disorders, such as mildew and aphids, may be evident now, especially in a dry or warm season. Watch out for weeds, too, and control these while they are still small. Carefully hoe round established plants on a warm, dry day to kill annual weed seedlings and use a trowel or border spade to dig up invasive perennial weeds, such as thistles, couch grass, horsetail and ground elder.

Avoiding propagation problems Fluctuating temperatures
mean that early spring is a fickle season under glass, and ventilation becomes increasingly important as plants start growing from seed or cuttings. The soft new growth is very susceptible to pests and diseases, and you need to be sensitive to the needs of these young plants to avoid common problems.

Powdery mildew affects leaves (above right) and stems (above left) in a warm, dry spring following a mild winter that allowed dormant spores to survive. Cut out affected stems and spray with fungicide.

Half-hardy annuals grown under glass are prone to damping-off disease. Here, a whole tray of antirrhinums has been wipe out. Reduce the risk of infection by using clean equipment and fresh compost and watering with a copper-based fungicide.

Damping-off disease

This very troublesome disease can result in the death of many kinds of seedlings, indoors and out. It is caused by a range of virulent soil-borne fungi, which are liable to affect fast-growing annual bedding flowers and some vegetables while they are very young.

Seedlings develop dark leaves and then collapse in patches, their white stems become as thin as hairs, turn brown and shrivel at soil level. The condition can quickly affect a whole tray or bed of seedlings. Treatment is difficult and control depends on precautionary good hygiene.

■ **Use fresh seed compost** and sterilise any garden soil used for compost mixes.

■ **Wash and disinfect old pots** and plant labels before reusing them.

■ **Mix horticultural sand** into compost (left) to improve drainage for susceptible plants, such as nemesia and antirrhinums, and cover seeds with a thin layer of sand.

■ **Sow sparsely** and prick out or thin before the seedlings are overcrowded.

■ **Use modular trays**, sowing a few seeds in each cell, to confine infection, which spreads easily throughout ordinary seed trays.

■ **Water containers** from below using mains water, as rainwater from butts is often a source of the fungi. Avoid overwatering.

■ **Water with a copper-based** fungicide after sowing and pricking out.

■ **Keep seedlings at the right temperature,** ideally in a heated propagator: cold draughts and cool, wet soils or composts are lethal.

Sowing in individual pots or modular trays will reduce the risk of disease spreading.

Sterilising soil for compost

It is very important to sterilise garden soil to kill most of the disease pathogens within it before you add it to your homemade compost mixes. You can use a chemical disinfectant, available specially for this purpose, or you can pasteurise the soil in a conventional or microwave oven following the instructions given below.

■ **In a conventional oven** Spread moist soil 8–10cm deep in a tray, cover with kitchen foil and heat at 80ºC (175ºF) for 40 minutes. Uncover and leave to cool.

■ **In a microwave oven** Sieve out the stones and organic material. Weigh 2kg of soil into a bowl, cover and cook on maximum for five minutes. Spread on a tray to cool.

■ **Do not sterilise** garden compost or leaf-mould, as heat can reduce their food value.

Jobs for late spring

Late spring

What you do in the garden in late spring will lay the foundations for the summer and will go a long way to determining the success of your garden in the remaining seasons.

■ **Prick out seedlings** under glass as soon as they are large enough to handle, and thin or transplant outdoor sowings before they become overcrowded.

■ **Harden off plants** ready to go out into the open garden and feed any that are left waiting beyond their transplanting date.

■ **Start a weed control routine**, by hand-pulling, hoeing or spraying as appropriate.

■ **Water and ventilate** the greenhouse freely, but keep an eye on the possibility of night frosts.

■ **Water outdoor plants** regularly in dry weather, and mulch to conserve soil moisture.

■ **Look for early signs of pests** and diseases, especially on fruit.

Feeding routines

ERICA ARBOREA

After winter and early spring rains, soils will be depleted, and many plants benefit from feeding now when they are actively growing.

■ **Give spring-flowering bulbs** a single feed of general fertiliser when their flowers fade, to help them with food storage for next season.

■ **Give acid-loving plants** such as rhododendrons, azaleas, camellias, pieris, heathers (Erica) and tree heathers (E. arborea, above) a dressing of special ericaceous fertiliser.

■ **Give a general granular feed** to all other shrubs, fruit, herbaceous perennials, and hedges, especially those cut back hard to restore their shape. Treat these feeds as a spring tonic to supplement long-term food supplies that come from dressings of well-rotted manure or garden compost forked in or applied as a mulch.

Rake gravel drives and paths, and spot-treat any perennial weeds with systemic herbicide on a still, warm day. Brush down steps, paved paths, decking and patio surfaces, and treat them with an algicide if they are green and slippery.

Tidying the garden
Keeping the garden tidy is not just cosmetic. Even the wildest natural garden needs maintenance to prevent unwanted plants from taking over, and to discourage pests and disease. Formal gardens need more routine tidying, and you must decide how much orderliness is appropriate or manageable.

■ **Clear away** dead topgrowth in borders left for plant protection or to help wildlife, cutting down to ground level. Pull up spring bedding and discard or compost the waste.

■ **Fork lightly** over beds and borders to remove weeds and self-sown seedlings. This will also break up any surface crust and expose soil pests to foraging birds. Fork in decaying mulches, and lightly loosen newer mulches that might be effective for another season so that rain can penetrate. Take care not to damage plant roots.

■ **Collect fallen branches**, twigs and leaves left over from autumn. Clear all pruning debris and dead material from frost-damaged and injured plants. Provided they are not diseased, you can shred these and add them to a compost heap.

Weed patrol
Controlling weeds is a necessary task at any season, but at this time of year they seem to appear almost anywhere. Dormant seeds in freshly disturbed soil germinate quickly.

■ **Identify weeds** carefully when tidying beds and borders, as useful flower, shrub and tree seedlings often start appearing now, especially after a cold winter.

■ **Hoe bare ground** regularly to prevent weeds from getting beyond the seedling stage – this is particularly important in vegetable beds, since many weeds host the same pests and diseases as cultivated crops.

■ **Mulch** wherever possible to smother and suppress weed germination as your soil warms up. Pull larger weeds by hand, but stubborn perennial species and those growing in gravel, paths and other hard surfaces may need spot treatment with a systemic herbicide to prevent regrowth.

Continuing protection
Spring is the most fickle season, with sharp frosts and sudden cold snaps that can surprise unprepared gardeners, as well as periods of warm sunshine.

■ **Do not hurry** to unwrap pots and containers insulated last year against frost.

■ **Keep cloches** and fleece handy to cover outdoor sowings and recent transplants.

■ **Remember to harden off** young plants raised outdoors under glass or plastic to fully acclimatise them before planting them in their flowering positions.

Seasonal threats
Pests and diseases begin to appear as the temperature rises towards the end of spring, particularly if the weather is also damp. Keep a watchful eye for early signs of problems so that you can stop them taking hold.

■ **Powdery mildew** This disease thrives in

overcrowded and dry conditions and shows as a greyish white coating on leaves and shoots. Avoid by pruning and spacing plants to admit plenty of air, water plants well in dry weather, and remove any affected growth such as white, distorted shoots on apples or dying forget-me-not leaves.

■ **Aphids** These start multiplying in mid and late spring. Disperse them with a jet of water from a hose, spray with insecticide, or encourage their natural predators (page 12).

■ **Caterpillars** Various kinds appear on plants as soon as moths and butterflies are on the wing. Look under leaves for egg clusters, which can be crushed with finger and thumb. Remove caterpillars whenever you notice them.

■ **Slugs and snails** These all-year pests are particularly active in spring, feeding hungrily on young growth and seedlings. Collect them on damp evenings for disposal; trap them under orange or grapefruit skins laid on the ground; use proprietary beer traps; discourage them with barriers of grit or crushed eggshells; encourage frogs, toads and thrushes; or, as a last resort, use slug pellets sparingly, scattered thinly but frequently, especially in damp weather.

■ **Vine weevils** Both adult weevils and their underground grubs are a continuous threat to pot-grown plants and open ground plants such as bergenias and strawberries, but they are particularly active now. Water plants in containers with biological controls based on parasitic nematodes or imidacloprid. Keep the soil regularly cultivated round outdoor plants to expose the grubs to birds.

Fruit pests

Fruits are susceptible to a number of pests and ailments, often at very specific times during spring. If you are following a chemical spray routine, these are the critical stages for specific fruit. Do not spray when flowers are open.

■ **Strawberries** Spray insecticide against aphids from April until harvest.

■ **Raspberries** Use a systemic fungicide in April to prevent fungal diseases.

■ **Gooseberries** Spray with fungicide against mildew when flowers first open and again when berries are visible. Treat against caterpillars in May with insecticide.

■ **Blackcurrants** Control big bud mites by spraying insecticide when flower buds resemble miniature bunches of grapes, and again a month later.

■ **Apples and pears** Use a combined insecticide and fungicide while flower buds are green, and again when they show colour, to treat most common problems.

■ **Plums and cherries** Spray insecticide at green bud stage to control aphids and grubs.

Cabbage white caterpillars are a constant threat to cabbages and other brassicas.

Summer

Jobs for summer

Watering, weeding and deadheading in the garden throughout summer will keep plants in peak condition, strong enough to shrug off diseases, and helps you to spot pests at the earliest opportunity. In summer, pests and diseases can become established quickly, so keep your eyes open, act fast and consider taking preventative action of the organic kind (see pages 14–15) before you go away on holiday. Consider asking a friend or neighbour to care for your garden, and particularly any containers, while you are away.

■ **Prick out**, transplant or thin seedlings as soon as you can handle them, so that they have plenty of room to develop.

■ **Lift and dry** spring-flowering bulbs; tidy those that you leave in the ground.

■ **Keep the greenhouse cool** by shading, ventilating and damping down.

Coping with pests and diseases

This is the main growth or migration time for many pests and diseases, with conditions just right for their rapid establishment, so take every precaution to avoid or prevent trouble.

■ **Inspect plants regularly** for the first sign of problems; hand-pick caterpillars, and collect slugs and snails late in the evening after rain.

■ **Practise good hygiene** by clearing up fallen fruit, dying leaves and other plant debris that can harbour pests and disease.

■ **Use mechanical** and biological controls where possible (see page 37).

■ **If you need to spray**, use a chemical specifically targeted at a particular pest or disease; always follow the manufacturer's instructions exactly. Afterwards, wash your hands and face, and all equipment, and dispose of any leftover solution safely on a vacant, out-of-the-way patch of soil.

■ **Never spray in bright sunshine** or windy weather; avoid spraying open flowers or wet leaves, and do not spray when bees are around.

Watering

Treat watering as a top priority, especially for plants in containers. Outdoor plants can also have huge water requirements for steady growth, especially on light soils and in exposed gardens. On average, 2.5cm of rain every eight to ten days is needed to keep soils moist.

Annually working in organic material such as garden compost and well-rotted manure helps the soil to retain moisture, as does mulching the surface when the soil is moist. When you do water, give substantial and regular amounts rather than an occasional light sprinkling.

Leaf cutter bees leave a distinctive hole in foliage (above). They do not harm the plants, and will help with pollination.

Snails congregate (right) in empty containers and cool corners where plant debris has been left to rot.

Capillary matting draws up water from a reservoir into the compost in seed trays and pots, reducing the need to water manually.

■ **Water in dull weather**, in the early morning or during the evening.
■ **Concentrate on vulnerable plants:** leafy vegetables, flowering and maturing fruit, seedlings and plants that have been recently planted or moved, and those growing near walls or in containers.
■ **Direct water to the roots** of thirsty plants by burying pots or sections of drainpipe close by and filling these, or by creating a depression around the plant to hold water.
■ **Check plants in pots** and greenhouses at least once a day as the weather begins to get hotter and drier.
■ **Stand pots in trays** lined with capillary matting, immersing one end in a reservoir of water (see above). In this way plants are kept moist for several days.
■ **Collect rainwater** as a useful precaution if drought threatens.

Weeding
Many weeds grow fast in hot summer weather. If allowed to flower, they scatter thousands of seeds in a few weeks. It is important to keep weeds under control as they can attract pests that will quickly move on to your cultivated plants and will create overcrowded beds where diseases can flourish.

■ **Hoe bare soil regularly**, ideally while weed seedlings are small and the soil is dry.
■ **Pull up** annual weeds before they flower.
■ **Never leave weeds** lying on moist ground, where they may continue to ripen and seed.
■ **Fork up** perennial weeds while small or spot-treat with a systemic weedkiller.

Holiday tips
Before you go on holiday take precautions to prevent problems while you are away.
■ **Thoroughly water** everywhere the night before you leave if the weather is dry.
■ **Mulch drought-sensitive** plants after soaking them.
■ **Water and feed** greenhouse plants, and move them away from sunny windows to stand in self-watering containers or a bowl or sink of shallow water.
■ **Mow the lawn**.
■ **Deadhead all fading flowers** and remove dying leaves, especially from plants in pots.
■ **Gather ripe fruit and vegetables**, together with any that are nearly ready.
■ **Get up to date** with pricking out and potting on so that young plants continue growing unchecked.
■ **In the greenhouse** ensure there is adequate shading. An automatic ventilator and watering system is an expense, but you may consider it a worthwhile investment.
■ **Arrange for someone** to water the garden while you are away, and pick ripening fruit and vegetables in payment for their time.

Run the Dutch hoe blade just below the soil surface to cut through weed seedlings. Put them on the compost heap or burn them.

Late summer

Jobs for late summer

Use this checklist to make sure you have not overlooked any important seasonal jobs.

■ **Continue watering** as often as necessary, focusing on vulnerable plants and those in containers.

■ **Harvest vegetables** while they are in peak condition, and sow or plant follow-on crops or green manures.

■ **Pick fruit** as crops mature and protect ripening fruit from birds, squirrels and wasps.

■ **Be extra vigilant** for pests and diseases, which often flourish in the late summer heat and humidity.

■ **Monitor your greenhouse** to maintain a congenial level of heat and humidity through ventilation and regular damping down.

■ **Keep on top of weeds**, especially fast-growing annuals, which should be cleared before they flower and shed their seeds.

■ **Deadhead** flowering plants regularly to prolong their display.

■ **Thoroughly clean** the inside of the greenhouse before plants need rehousing in autumn.

■ **Lift and divide** overcrowded perennials late in the season.

■ **Check plant supports and ties** to make sure that they are secure before the strong autumn winds arrive.

Conserving water
You can reduce the impact of prolonged drought in late summer by following these simple steps.

■ **Top up mulches** applied earlier in the year.

■ **Continue controlling weeds**, which compete with other plants for soil moisture.

■ **Lay a semi-permeable membrane** over borders before planting and then mulch over the top of it. You can also cut sheets to lay them around existing planting. The matting will help to keep moisture within the soil.

■ **Install water butts** or other containers to collect rainwater from downpipes.

■ **Save 'grey' water** (from the bath and washing-up) for watering.

Planting through a semi-permeable membrane will conserve moisture in the soil and help to suppress weed growth, too.

Mulching the surface of the compost in containers helps to stop it from drying out.

Coping with pests and diseases

It is tempting to relax your vigilance once the main fruit and vegetable harvest begins and the end of the flowering season approaches. But late summer pests and diseases may lie dormant until next year if left untreated.

■ **Inspect plants carefully** as you work in the garden. Make a habit of checking shoots and the undersides of leaves looking for insect pests and spores or other signs of fungal infections.

■ **Late in the season** it is often more effective to cut off and burn leaves, shoots or flowerheads affected by disease or aphids, rather than attempt treatment with slow-acting systemic sprays.

■ **As temperatures drop under glass** so too does the efficacy of biological controls, most of which require consistent warmth for the predator or parasite to feed and multiply.

Seasonal threats

Keep a particularly close watch for these common late-summer problems.

■ **Brown rot** Soft brown patches develop on fruits such as apples, plums and pears. Prune off affected shoots, pick up and destroy windfalls, and do not store infected fruit. Remove shrivelled fruits from the tree.

■ **Fireblight** A bacterial disease causing shoots and leaves to wither and turn brown, as if scorched. Cotoneasters, apples, pears and Sorbus and their relatives are all vulnerable. Cut out affected shoots about 60cm below the affected area to clean the wood. Dip tools in garden disinfectant during and after use. Badly affected plants are better dug up and destroyed.

■ **Grey mould (botrytis)** Fluffy, off-white mould develops in damp conditions on the stems, flowers and leaves of most plants. Clear up all loose dead material from around plants. Pull up dying plants or cut off affected portions and destroy them. Improve ventilation under glass.

■ **Downy mildew** In damp, humid

Inspect shoots and leaves regularly so that you can catch signs of disease and infestation in the early stages.

conditions, yellowish patches develop on upper leaf surfaces with mealy white outgrowths on the undersides. Remove affected leaves, improve air circulation in the greenhouse by better spacing and ventilation, and grow resistant varieties.

■ **Powdery mildew** In dry conditions foliage and stems develop a greyish white, powdery coating that later turns brown. Pick off affected leaves, water and mulch in dry weather, and avoid using high-nitrogen fertilisers after the longest day. Some fruit varieties can be sprayed with sulphur.

■ **Cabbage whitefly** Tiny delta-shaped flies are easily disturbed from cabbage and other brassica crops. Remove lower or outer leaves, spray in the early morning with insecticidal soap, and burn or compost plants immediately after cropping.

■ **Caterpillars** These may still be a problem late in the season. Examine plants, especially the undersides of the leaves, when the adult moths and butterflies are seen on the wing. Crush the egg clusters and pick off the grubs by hand.

■ **Earwigs** A useful insect that feeds on aphids, but also damages flowers such as dahlias and chrysanthemums. Clear loose plant debris from around plants. Trap the earwigs in upturned pots stuffed with straw, and empty them well away from the affected plants.

■ **Red spider mites** The minute mites suck sap from the leaves, which turn pale and mottled. Fine webs may be visible on badly affected plants. Introduce the predatory Phytoseiulus mite under glass up to the end of August, and raise humidity levels by damping down. Outdoors, spray plants with insecticidal soap.

Autumn

Jobs for autumn There
are many clearing up and cultivation tasks that need doing before winter, all of them essential to maintain the good health of your plants.

■ **Start removing dead** plant remains from borders and vegetable beds for composting.

■ **Rake up leaves**, particularly those fallen on alpines and herbs which cannot stand damp conditions, and stack for leaf-mould.

■ **Finish cleaning** the greenhouse, coldframes and cloches. Wash and dry pots and trays, and store neatly.

■ **Insulate the greenhouse**, and prepare plants inside for colder weather.

■ **Move outdoor containers** under cover or insulate them where they stand.

■ **Drain and roll up** hosepipes and clean out water butts and watering cans. Hang

Clear dead and dying plant growth from borders and put it on the compost heap.

Colourful annuals, such as these variegated trailing nasturtiums, will continue to flower until the first frost causes them to collapse.

watering cans to dry to prevent algae forming inside.

■ **Continue to check plants** for problems, especially diseases.

Preparing for frost

Plan ahead so you are ready to protect plants if a sharp frost suddenly occurs after a long, warm autumn.

■ **Do not feed plants** with high-nitrogen fertiliser during autumn, as this encourages soft growth that will be vulnerable to injury.

■ **Choose hardy varieties** of winter vegetables and grow them in the warmest part of the garden.

■ **In very cold areas**, only grow fruit that ripens early, so that you can harvest it before conditions deteriorate.

■ **Mulch plants** of borderline hardiness with straw, leaves or crumpled newspapers held in place with wire netting.

■ **Shield young and tender shrubs** with windbreaks of plastic mesh or sacking.

■ **Do not tidy perennials** too thoroughly. Some dead growth left in the flower beds will help to insulate roots and trap a blanket covering of fallen leaves.

■ **Keep a supply of newspapers**, old curtains, blankets or fleece to cover coldframes, cloches and vulnerable plants outdoors.

■ **Check plants** after a hard frost, especially those that have been recently planted. Plants can be forced out of the soil as moisture in the ground freezes and expands and plants may need to be firmed in once more after a very cold spell.

Seasonal problems

Pest attacks diminish during autumn, but many diseases thrive in damp, mild conditions. There is still time to spray with fungicide if you notice symptoms of rust, mildew or black spot, but taking sensible precautions against the most common problems will have longer-term impact.

■ **Clear leaves** and other plant debris unless needed for frost protection.

■ **Gather fallen fruits** and put them out for the birds in a corner, well away from plants. Pick rotten or mummified fruits and remove or burn them.

■ **Part prune shrubs** or thin out congested stems to improve the air circulation.

■ **Ventilate plants** under glass whenever possible, and water only when absolutely necessary, preferably early in the day.

Pick mummified tree fruits as soon as you see them, but do not compost them.

Winter

Jobs for winter

While plants are dormant, take the opportunity to stand back and reassess your garden. Clean tools and other equipment, so they are ready for use in spring.

■ **Clear fallen leaves**, especially from hedge bottoms, ponds and alpines that might rot, and stack them where they can decay into leaf-mould. Don't clear leaves from wildlife hedges as they provide useful winter habitats for a range of animals and insects.

■ **Clean and store** pots, trays, canes and labels for spring.

■ **Shake settled snow off** evergreen hedges and shrubs.

■ **Keep fleece and bubble plastic handy** to protect outdoor containers, coldframes and plants in unheated greenhouses.

■ **Empty compost heaps** when the contents have decayed sufficiently, and spread on bare ground as a mulch or for digging in. Spread rotted manure in the same way.

■ **Turn off and drain** outside taps and empty and clean water butts in case the water freezes.

Clean secateurs with a wire brush (above) to remove any accumulation of sap then use a metal file (right) or sharpening stone to sharpen the blades. Hold the sharpening tool at the angle of the blade edge.

■ **Continue cultivating heavy soil**, but leave light soils until spring before digging; cover with compost or manure to protect the surface.

■ **Clean paths**, patios and other hard surfaces if necessary, pressure washing or scrubbing them with an algicide to prevent surfaces becoming green and slippery.

■ **Keep off the lawn** whenever it is wet or frosted.

Taking care of tools

Take the opportunity to clean and maintain pruning and other tools at this quiet time of year. It is vital to keep tools scrupulously clean in order to avoid spreading disease.

■ **Scrape off accumulated soil** with a paint scraper or offcut of wood, and finish off with a wire brush.

■ **Sharpen the blades** of spades and hoes with a file, and hammer out dents on a flat surface.

■ **Clean secateurs**, loppers and shears with a wire brush (above). If necessary, dismantle the parts and clean individually. Sharpen the cutting edge with a file before storing.

■ **Repair broken handles** and paint all

Lag or insulate outside taps and any water supply pipes than run outside the house.

wooden parts with linseed oil; spray metal parts with oil or rub them over with an oily cloth before hanging them up.

■ **Rinse out sprayers** with warm soapy water; dismantle and clean all parts. Smear a little petroleum jelly on all sealing rings before you reassemble.

Using windbreaks

Gardeners rightly regard frost as a serious threat to many plants, but high winds can be just as damaging, breaking stems and shoots, and disturbing roots, all leaving plants with wounds through which diseases can enter. A solid fence or wall aggravates this effect, whereas a hedge or line of shrubs or trees filters wind and shields plants for a distance of about five times its height.

Shelter new plants in exposed gardens with a temporary screen of windbreak netting, wattle hurdles or open fencing until they are established. Check after gales that the screen is secure and undamaged. For a more permanent windbreak, plant a boundary hedge or shorter hedges within the garden.

Looking after birds

Many birds are useful allies in your efforts to control plant pests. Throughout winter, they will appreciate a regular supply of food, either on a bird table safe from cats and squirrels, or in feeders suspended among branches. Mould some balls of fat mixed with seeds or hang birdcake logs (below). Keep bird feeders full of concentrated foods, such as black sunflower seeds, peanuts and nigella seeds. Hang several feeders around the garden, and move them occasionally to encourage shy birds.

Introduce bird boxes to your garden to provide shelter over winter and you may encourage the birds to stay when nesting season begins in spring. A regular supply of fresh water is also vital, especially when their usual supplies may be frozen.

Making a birdcake log

1 Soften a block of lard to a creamy consistency then stir in an equal volume of rolled oats and add any of the following: sunflower seeds, finely chopped nuts, sesame seeds, linseeds or leftover seeds from your health store. Quantities can vary – there is no need to measure.

2 Split a narrow log in half and chisel out several holes on the rounded side. Pack in the mixture firmly, finishing flush with the surface.

3 Hang the split log against a wall to feed birds such as nuthatches and woodpeckers.

Directory of pests & diseases

32 Pests and diseases

THE FOLLOWING PAGES ARE A GUIDE TO THE MOST COMMON GARDEN PESTS AND PLANT DISORDERS AND THE BEST METHODS OF TREATING THEM, INCLUDING ORGANIC AND BIOLOGICAL REMEDIES WHERE POSSIBLE.

The list of approved chemicals and the brand names of the products which contain them is constantly under review by the Department for the Environment, Food and Rural Affairs (DEFRA) and the availability of some products recommended here may change. Even when trade names change, the chemical contents are always given on the label. It is always advisable to consult your local garden centre for advice about appropriate approved treatments.

A number of pests and diseases in the list are annotated as 'Notifiable', meaning that their presence in your garden must be notified to DEFRA.

Ladybirds, birds and other predators of insects are the gardener's allies. Refrain from spraying insecticides at the first sign of attack to give these natural armies a chance to go to work. In many cases, pests, diseases and plant disorders can be controlled with good growing techniques and garden hygiene, so that you need only use chemical sprays on those plants that are most seriously affected.

Treatments and their trade names

ALUMINIUM SULPHATE * Fertosan slug and snail powder, Growing success, Doff Slug Attack
AMBLYSEIUS SPECIES ** Thrip control
ANAGRUS ATOMUS ** Leafhopper parasite
APHIDUS SPECIES ** Aphid parasite
APHIDOLETES ** Aphid parasite
BENDIOCARB Doff Wood Lice Killer
BIFENTHRIN Doff Garden Pest Killer, Bayer Bug-free, Bayer Sprayday Greenfly Killer, Scotts Bug Clear, Scotts Rose Clear (with mycobutonil)
CHELATED COMPOUNDS Chelated Trace Element Mix, Miracle Miracid, Murphy Sequestrene
COPPER OXYCHLORIDE * Murphy Traditional Copper Fungicide, Copper Tape, Snail Tape, Slug and Snail Rings
COPPER SILICATE Doff Socusil Slug Repellant
COPPER SULPHATE Bayer Cheshunt Compound (with ammonium carbonate)
CRYPTOLAEMUS ** Mealy bug predator
DELTAMETHRIN Doff New Ant Killer Spray, Bayer Ant Killer Spray
ENCARSIA FORMOSA ** Whitefly predator
FATTY ACIDS * Bayer Organic Pest Control, Doff Greenfly and Blackfly Killer, BabyBio House Plant Insecticide, Green Fingers Organic Pest Spray
FERROUS SULPHATE Phostrogen Moss Killer and Lawn Tonic
FRITTED TRACE ELEMENTS Fritted Trace Elements
GREASE BANDS * Bayer Boltac Greasebands, Vitax Fruit Tree Grease, Agralan Glue Bands
HETEROHABDITIS MEGIDIS ** Chafer Bug Predator
HYPOASPIS ** Sciarid predator
IMIDACLOPRID Bayer Provado Ultimate Bug Killer concentrate (with sunflower oil), Bayer Provado Ultimate Bug Killer ready to use, Bayer Provado Ultimate Bug Killer aerosol (with methiocarb), Bayer Provado Lawn Grub Killer
INSECTICIDAL SOAP * see Fatty Acids
MANCOZEB Bayer Dithane 945
METALDEHYDE Bio Slug Mini-Pellets, Doff Slugoids Slug Killer, Slug Clear!
METAPHYCUS HELVOLUS ** Scale parasite
METHIOCARB Provado Ultimate Bug Killer aerosol (with imidacloprid)

MYCLOBUTANIL Doff Systemic Fungus Control, Bayer Multiuse, Bayer Fungus Fighter, Scotts Rose Clear 3
PENCONAZOLE Scotts Fungus Clear
PERMETHRIN Bio Kybosh (with pyrethrins), Bayer Ant Killer Dust, Doff New Ant Killer, Doff Foaming Wasp Nest Destroyer (with tetra-methin)
PHASMARHABDITIS HERMAPHRODITA ** Nemaslug
PHENOLIC EMULSION Armillatox
PHEROMONE TRAPS ** Oecas Pagoda Codling Moth Trap, Trappit Codling Moth Trap
PHYTOSEIULUS PERSIMILIS ** Red Spider Mite predator
PYRETHRUM AND PYRETHINS * Nature's Answer Natural Bug Killer, Doff All-in-one Bugspray
ROTENONE * Bio Liquid Derris, Doff Derris Dust, Vitax Derris Dust
STEINERNEMA FELTIAE ** Leatherjacket predator
STEINERNEMA KRAUSSEI ** Nemasys Vine Weevil predator
SULPHUR Nature's Answer Natural Fungus and Bug Killer
TAR OIL Jeyes Fluid
TETRAMETHRIN Doff Flying Insect and Crawling Insect Killer (with d-phenothrin)
THIACLOPRID Provado Vine Weevil Killer 2
TYPHLODROMUS PYRI ** Typhlodromus
YELLOW STICKY TRAPS * For greenhouse pest control

---

* Organic (or suitable for use by organic gardeners)
** Biological control agent

Aphids

Aphids

■ **PLANTS AT RISK** Most cultivated plants growing in the open, under glass or indoors.

■ **RECOGNITION** Colonies of 1–7mm long, round-bodied insects suck the sap from leaves and distort plant growth. They can spread viruses and excrete honeydew on which sooty mould grows. Aphids, which are mostly wingless, may be black, green, pink, red, yellow or variously coloured.

■ **DANGER PERIOD** Spring and early summer in the open, but any time of the year under glass or indoors.

■ **TREATMENT** Outdoors or in a greenhouse, spray thoroughly with systemic insecticide such as imidacloprid, or with non-systemic insecticides such as bifenthrin, fatty acid, pyrethins or derris. Inside the house, use derris only. **See also** WATER LILY APHIDS (p. 49). **Organic advice** Encourage natural predators – ladybirds, bluetits, hoverflies, beetles (see p. 12).

Asparagus beetles

■ **PLANTS AT RISK** Asparagus.

■ **RECOGNITION** Yellow-and-black, 6mm long beetles and their humpbacked larvae feed on the foliage and shoots. A severe infestation of the beetles can strip plants.

■ **DANGER PERIOD** May and early June.

■ **TREATMENT** Spray with derris or bifenthrin as soon as damage is seen. **Organic advice** Pick off and destroy beetles as they are seen. Clean up plant debris at all times, but especially in late autumn to prevent overwintering.

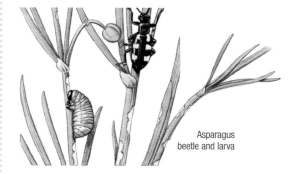

Asparagus beetle and larva

Cabbage whiteflies
▶ **SEE** page 106

Caterpillars
▶ **SEE** page 40

Cutworms
▶ **SEE** page 74

Flea beetles
▶ **SEE** page 40

Froghopper nymph

Froghoppers

■ **PLANTS AT RISK** Dendranthemas, roses, lavenders, solidagos, perennial asters and many other plants.

■ **RECOGNITION** Frothy masses of 'cuckoo spit' on stems and leaves. Each patch of froth conceals a green, sap-feeding froghopper nymph, about 5mm long.

■ **DANGER PERIOD** May to July.

■ **TREATMENT** Spray forcefully with bifenthrin or permethrin, or a systemic product. **Organic advice** Damage is seldom severe enough to warrant control. A strong jet of water should dislodge the creatures.

Glasshouse whiteflies

■ **PLANTS AT RISK** In a greenhouse: tomatoes, cucumbers, fuchsias, abutilons, dendranthemas. In the house: coleus and pelargoniums.

■ **RECOGNITION** The 1.5mm long whiteflies and scale-like larvae are found on the underside of leaves, which become soiled with honeydew and sooty mould.

■ **DANGER PERIOD** All year.

■ **TREATMENT** In a greenhouse, spray regularly with bifenthrin, fatty acids or permethrin or use a systemic, such as imidacloprid. In the house, use imidacloprid. **Organic advice** Suck up the flies with a vacuum cleaner. Hang up sticky traps. Introduce *Encarsia formosa*, a parasitic wasp, as soon as pests are seen. Spray plants with insecticidal soap.

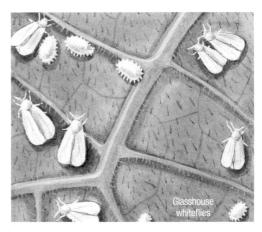

Glasshouse whiteflies

Greenhouse pests IN THE WARM CONDITIONS OF A GREENHOUSE, NUMEROUS PESTS AND DISEASES CAN MULTIPLY RAPIDLY, so stay alert for early signs and take prompt action. Simple precautionary measures include keeping the atmosphere humid to suppress red spider mites and hanging up sticky yellow cards (below) as traps for flying insects such as whitefly.

Aphids, whitefly, scale insects, red spider mites and vine weevils are the commonest greenhouse pests. They can be kept in check by introducing a specific parasite or predator.

Release the biological control agent when you see the first signs of the pest in early summer. Remember to refrain from using conventional sprays because the insecticides will kill the biological control agent.

Biological control methods

TO CONTROL:	YOU NEED:
▪ aphids	▪ *Aphidius matricariae*
▪ red spider mites	▪ *Phytoseiulus persimilis*
▪ scale insects	▪ *Metaphycus helvolus*
▪ vine weevils	▪ *Heterorhabditis megidis*
▪ whiteflies	▪ *Encarsia formosa*

Gypsy moth caterpillars

Gypsy moths

■ **PLANTS AT RISK** Wide range of shrubs and trees.
■ **RECOGNITION** Infestations of hairy caterpillars, up to 6cm long. They are usually brownish yellow marked with black and with hair sprouting from wart-like bumps along their sides and backs. The ten on the back closest to the head are blue, others on the back are reddish brown and the rest are a yellowish colour. The hairs cause an allergic reaction in some people, so handle the caterpillars with care.
■ **DANGER PERIOD** April to July.
■ **TREATMENT** Notifiable.

Lily beetles

■ **PLANTS AT RISK** Lilies, fritillaria, nomocharis and polygonatum.
■ **RECOGNITION** Scarlet beetles, 8mm long, with black heads and legs feed on the leaves and other aerial parts of plants. Primarily found in parts of Surrey.

Lily beetle

■ **DANGER PERIOD** May onwards.
■ **TREATMENT** Spray with derris or imidacloprid. **Organic advice** Check plants frequently in spring and remove any beetles. Spray plants with derris.

Mealy bugs

■ **PLANTS AT RISK** Greenhouse and house-plants, especially orchids, succulents, citrus, hippeastrums, camellias and vines.
■ **RECOGNITION** Patches on leaves and stems of sap-eating pink insects, up to 4mm long, covered with woolly or mealy white wax.
■ **DANGER PERIOD** Any time.

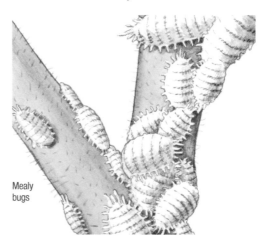

Mealy bugs

■ **TREATMENT** Use a systemic insecticide such as imidacloprid or a non-systemic insecticide such as bifenthrin. **Organic advice** Introduce the biological control agent *Cryptolaemus montrouzieri* – a beetle that eats mealy bugs.

Scale insects

■ **PLANTS AT RISK** Particularly troublesome on greenhouse and houseplants, but also ornamental shrubs, trees and fruit grown outside.
■ **RECOGNITION** Brown, yellow or white scales – flat or oval and 1–6mm long – found mainly on the underside of leaves and clustered along the veins and on the stems.

■ **DANGER PERIOD** Late spring or early summer outdoors, but at any time of the year when under glass.

■ **TREATMENT** Spray with fatty acids and permethrin, three times at two-week intervals or use imidacloprid. **Organic advice** A parasitic wasp, *Metaphycus helvolus*, can be bought for use in a green-house. Alternatively, gently remove scales by hand or with a soft toothbrush, or spray with insecticidal soap.

Scale insects

Biological controls

INSTEAD OF CHEMICAL TREATMENTS DESIGNED TO ERADICATE GARDEN PESTS, YOU CAN USE BIOLOGICAL CONTROL AGENTS WHICH REDUCE THE NUMBER OF PESTS TO AN ACCEPTABLE LEVEL WITHOUT UPSETTING THE BALANCE OF NATURE. Their actions are very specific and they are harmless to other creatures. Because the control agents cannot survive without the pests on which they feed, there is no danger that today's predators will become tomorrow's pests.

Biological control agents combat pests in various ways. Natural predators hunt down the pest species for food. For example, the beetle *Cryptolaemus montrouzieri* preys on mealy bugs. Parasitic agents lay their eggs in the eggs or larvae of the target pest. The parasitic wasp *Encarsia formosa* lays its eggs in whitefly larvae. As they develop the parasites eat their hosts from the inside, later emerging as adult wasps to carry on with the good work.

If slugs or vine weevils are destroying your garden, send for the nematodes (parasitic worms), available from garden centres. Unlike some non-biological remedies, such as slug pellets, these tiny worms will act without endangering the lives of hedgehogs or birds. They should be used only when the soil temperature is above 15°C.

Other biological control agents work by causing disease in their host. An example is *Bacillus thuringiensis* – a deadly bacterium that infects caterpillars. Apply this biological control agent as a spray solution to brassicas.

Lastly, synthetic imitations of insect pheromones can be used to lure male codling and plum moths into a sticky trap. The traps are available from garden centres. Hang them in fruit trees in May.

Biological control agents The agents are available by mail order. Biological control cards can be bought at garden centres and sent to the appropriate supplier as soon as a pest becomes apparent. Some agents are for use in greenhouses, others are for use outdoors. For more biological control agents that can be used in the greenhouse see p. 35.

PEST	CONTROL AGENT	TYPE
Aphids	Aphidoletes aphidimyza	Predatory midge larva
Caterpillars	Bacillus thuringiensis	Bacterium*
Codling moths	Pheromone	Trap*
Flower thrips	Amblyseius cucumeris	Predatory mite
Leaf miners	Dacnusa sibirica	Parasitic wasp
	Diglyphus isaea	Parasitic wasp
Mealy bugs	Cryptolaemus montrouzieri	Predatory beetle
Plum moths	Pheromone	Trap*
Sciarid flies	Hypoaspis miles	Nematode
Slugs	Phasmarhabditis hermaphrodita	Nematode*
Vine weevils	Steinernema carpocapsae	Nematode*
Whiteflies	Encarsia formosa	Parasitic wasp

(* SUITABLE FOR OUTDOOR USE)

Rhododendron leafhoppers

- **PLANTS AT RISK** Rhododendrons.
- **RECOGNITION** Dark green insects, 8mm long, with red stripes lay their eggs in the flower buds. Leafhoppers are harmless but leave entry sites for rhododendron bud blast (p. 81).
- **DANGER PERIOD** Late July to October.
- **TREATMENT** Spray with heptenophos & permethrin or fenitrothion two or three times in August and September.

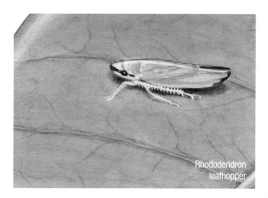

Rhododendron leafhopper

Solomon's-seal sawflies

- **PLANTS AT RISK** Solomon's seal and other polygonatums.

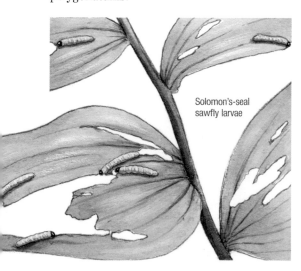

Solomon's-seal sawfly larvae

- **RECOGNITION** Grey-blue larvae, 2cm long, with black heads strip the foliage.
- **DANGER PERIOD** Summer.
- **TREATMENT** Spray with heptenophos & permethrin or malathion when the larvae are first seen. **Organic advice** Pick off the larvae when they begin to appear in late spring.

Weevils

▶ **SEE ALSO** VINE WEEVILS (p. 95)

- **PLANTS AT RISK** Various species of cultivated plants.
- **RECOGNITION** Adult weevils are about 9mm long, usually dark coloured and with a long snout; larvae are white legless grubs, about 1cm long, with a discernible light brown head. Both feed on roots, tubers, corms, stems, leaves, flowers and fruit, but damage is seldom severe.
- **DANGER PERIOD** Late spring, early summer.
- **TREATMENT** Dust or spray foliage, soil or potting composts with pirimiphos-methyl.

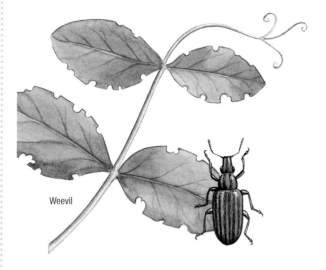

Weevil

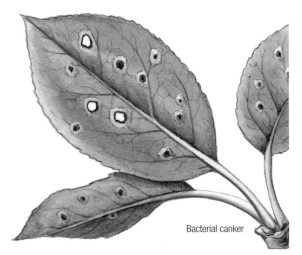

Bacterial canker

Bacterial canker

■ **PLANTS AT RISK** Primarily plums
(particularly 'Victoria'), but also peaches,
damsons and cherries, as well as
ornamental prunus.

■ **RECOGNITION** In late spring, small brown
spots appear on leaves. The spots later
drop out to give a 'shot hole' appearance.
Elongated lesions exude amber-coloured
gum on dying shoots.

■ **DANGER PERIOD** Autumn and winter,
although symptoms do not appear until the
following year.

■ **TREATMENT** Cut out and burn all infected
areas. Spray foliage with Bordeaux mixture
in mid August, mid September and mid
October. Prune during summer. **Organic
advice** Take care not to wound trees when
staking, tying up or strimming as this
leaves them open to infection.

Capsid bugs

■ **PLANTS AT RISK** Apples, beans, currants,
buddleja, dahlias, forsythias, hydrangeas
and many other plants.

■ **RECOGNITION** Tattered holes appear in
younger leaves as adult bugs and nymphs
feed on the sap. Flower buds and shoots
may be killed or deformed. On apples,
raised brown patches occur on buds, fruit
and leaves.

■ **DANGER PERIOD** April to August.

■ **TREATMENT** The bugs are 6mm long,
but have usually left before symptoms
are seen. Spray plants with heptenophos,
permethrin, fenitrothion or malathion
in spring, summer and early autumn.
In winter, clear up all garden debris.

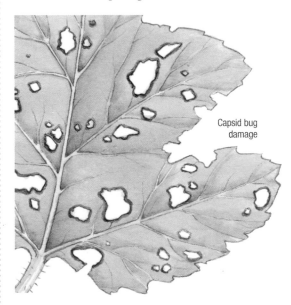

Capsid bug
damage

Caterpillars

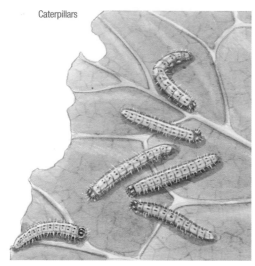

Flea beetles

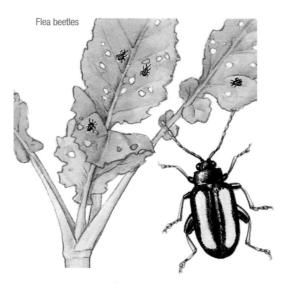

Caterpillars

- **PLANTS AT RISK** Many different species.
- **RECOGNITION** Foliage with irregularly shaped edges and holes. Caterpillars present.
- **DANGER PERIOD** From March onwards in the open but any time under glass.
- **TREATMENT** Crush small outbreaks by hand, including eggs. Spray large infestations with bifenthrin, pyrethin and permethrin, or derris as soon as the buds open. Wrap a grease band around trees in mid autumn to trap wingless female moths on apple trees. **See also** TORTRIX CATERPILLARS (p. 49). **Organic advice** Encourage wasps and birds.

Flea beetles

- **PLANTS AT RISK** Cabbages, radishes, turnips, wallflowers and related plants. Young seedlings are particularly at risk, especially in dry weather.
- **RECOGNITION** Young leaves become pitted with many minute holes by 3mm long shiny black beetles which jump when disturbed.
- **DANGER PERIOD** Sunny spells in April and May or when soil is dry.
- **TREATMENT** Dust vulnerable seedlings with derris. Practising good garden hygiene will help to reduce the risk of attack. **Organic advice** Encourage quick and early growth of plants. Water plants regularly in hot, dry weather. Grow highly susceptible plants under fine netting or horticultural fleece.

Gooseberry sawflies

- **PLANTS AT RISK** Gooseberries.
- **RECOGNITION** Leaf tissue stripped down to skeleton of veins by 2cm long, green, black-spotted caterpillars.

Gooseberry sawfly caterpillar

■ **DANGER PERIOD** April to August.
■ **TREATMENT** Spray early in May, or when symptoms first appear, with pyrethins or derris. **Organic advice** Frequently inspect the underside of the leaves at the centre of the bush, and pick off sawfly eggs and caterpillars. Cordons or espaliers are easier to treat.

Leaf-cutter bees

■ **PLANTS AT RISK** Many ornamental plants including laburnums, lilacs, privet and especially roses.
■ **RECOGNITION** Regular, roughly semi-circular pieces cut out of leaf edges.
■ **DANGER PERIOD** June and July.
■ **TREATMENT** Leaf-cutter bees are useful pollinators and in most cases should be tolerated. If control is essential, swat bees to remove them from the plants.

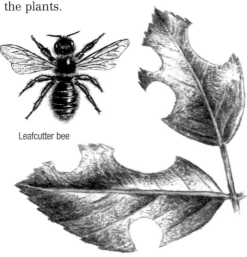

Leafcutter bee

Mustard beetles

■ **PLANTS AT RISK** Mustard, watercress, cabbages, swedes and turnips.
■ **RECOGNITION** Stems and leaves attacked by tiny metallic-blue beetle and its brown-yellow larvae.
■ **DANGER PERIOD** May to August.
■ **TREATMENT** Dust or spray with derris, but not on watercress, to avoid killing fish.

Raise the water level of watercress beds to drown beetles and larvae. **Organic advice** **See** FLEA BEETLES (p. 40).

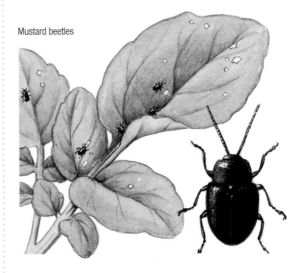

Mustard beetles

Rose slugworms

■ **PLANTS AT RISK** Roses.
■ **RECOGNITION** Holes eaten partly through leaf tissues, leaving transparent membranes.
■ **DANGER PERIOD** Symptoms may appear between May and September.
■ **TREATMENT** Pick off affected leaves as soon as they appear. Spray thoroughly with insecticidal soap, permethrin or derris. In winter, cultivate soil under affected plants to expose pupae to predators.

Rose slugworm

Shothole

■ **PLANTS AT RISK** Cherries, plums, peaches and ornamental prunus species.
■ **RECOGNITION** Leaves develop brown patches which become irregularly shaped holes, either due to leaf-spotting fungi or BACTERIAL CANKER (p. 39).
■ **DANGER PERIOD** Growing season.
■ **TREATMENT** Feed trees annually. Mulch and water well in spring, and give small trees a foliar feed. If symptoms appear next season, spray with copper fungicide during summer, and at leaf-fall.

Shothole

Slugs and snails

▶ **SEE** page 66

Solomon's-seal sawflies

▶ **SEE** page 38

Tortrix caterpillars

▶ **SEE** page 49

Vine weevils

▶ **SEE** page 95

Weevils

▶ **SEE** page 38

Water lily beetles

■ **PLANTS AT RISK** Water lilies, especially *Nuphar*.
■ **RECOGNITION** Holes are eaten in the upper surface of the leaves by the 6–8mm long, brown beetles and their 9mm long, dark brown or black larvae. The flowers may also be eaten.
■ **DANGER PERIOD** Early to late summer.
■ **TREATMENT** Spray plants with a strong jet of water to dislodge the beetles and larvae. Alternatively, submerge foliage with sacking or weighted netting for several hours.

Water lily beetle and larva

Anthracnose of trees

■ **PLANTS AT RISK** Weeping willows and London planes.

■ **RECOGNITION** Leaves curl, become discoloured and fall prematurely. Severe outbreaks can denude trees by August, followed by die-back.

■ **DANGER PERIOD** Spring or during mild wet summers.

■ **TREATMENT** Cut out infected shoots and burn, together with any fallen leaves. As leaves unfold spray small trees with copper fungicide, such as Bordeaux mixture, and repeat twice during the summer.

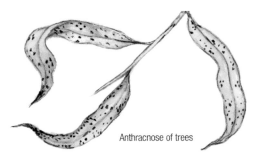

Anthracnose of trees

Box suckers

■ **PLANTS AT RISK** Box shrubs and trees.

■ **RECOGNITION** The tips of leaves curve inwards and form tight clusters. Caused by tiny sap-sucking yellow-green nymphs.

■ **DANGER PERIOD** Spring.

■ **TREATMENT** Spray in spring with a systemic insecticide, such as heptenophos & permethrin. Kill the adults in mid summer with imidacloprid spray. Cut out damaged shoots and destroy them.

Box suckers

Bulb scale mites

■ **PLANTS AT RISK** Narcissi and hippeastrums.

■ **RECOGNITION** The foliage of narcissi becomes slightly distorted. Cutting the bulb in half reveals brown marks. On hippeastrums, the leaves are malformed and stained with russet-brown scars.

■ **DANGER PERIOD** January to April on narcissi, especially if forced; any time on hippeastrums under glass.

■ **TREATMENT** Powdered sulphur may relieve the symptoms, but it is best to destroy badly affected plants. Maintain good hygiene in a greenhouse, and do not handle healthy bulbs after diseased ones. Expose dormant bulbs to frost for several nights before planting.

Bulb scale mites

Carnation ringspot
▶ **SEE** page 52

Celery leaf miners
▶ **SEE** page 52

Celery leaf spot
▶ **SEE** page 53

Cherry blackflies

Cherry blackflies

■ **PLANTS AT RISK** Flowering and fruiting cherries.

■ **RECOGNITION** Young leaves curl and twist as black aphids feed on them. Sticky honeydew and sooty mould often present.

■ **DANGER PERIOD** May to July.

■ **TREATMENT** Immediately after flowering, spray with a systemic insecticide, such as imidacloprid. In late December/early January treat with a tar-oil winter wash.
Organic advice Encourage the aphids' natural enemies – ladybirds, spiders, earwigs, hoverflies, parasitic wasps, ground beetles and bluetits (p. 12).

Corky scab of cacti

■ **PLANTS AT RISK** Cacti, especially opuntias and epiphyllums.

■ **RECOGNITION** Sunken patches of irregular rusty or corky scabs.

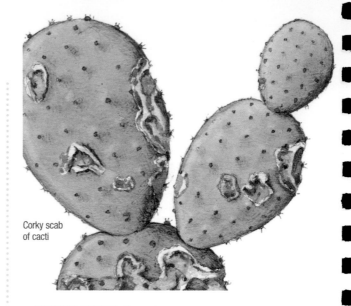

Corky scab of cacti

■ **DANGER PERIOD** Growing season.

■ **TREATMENT** Destroy badly affected plants and improve cultural conditions.

Crinkle
▶ **SEE** page 54

Cucumber mosaic virus

■ **PLANTS AT RISK** Cucumbers and many other plants.

■ **RECOGNITION** Leaves and fruit become mottled and puckered, and growth is stunted. Caused by an aphid-borne virus.

■ **DANGER PERIOD** Growing season.

■ **TREATMENT** Burn affected plants and control aphids. If the disease is a regular problem, grow resistant varieties.

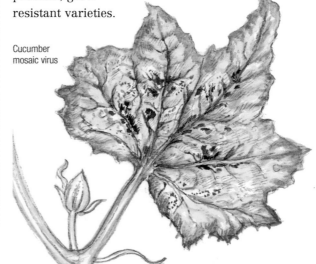

Cucumber mosaic virus

Eelworms

▶ **SEE ALSO:**
CHRYSANTHEMUM EELWORMS (p. 82),
POTATO-CYST EELWORMS (p. 118),
STEM AND BULB EELWORMS (p. 119)

■ **PLANTS AT RISK** Many herbaceous plants, vegetable crops and strawberries.

■ **RECOGNITION** Deformed and discoloured leaves or flowers caused by microscopic worms tunnelling through plant tissue.

■ **DANGER PERIOD** Growing season.

■ **TREATMENT** None. Destroy all infested plants. Remove all weeds, especially over winter. Remove all plant debris in autumn. Do not plant soft bulbs.

Eelworms

Frost damage
▶ **SEE** page 74

Gladiolus
▶ **SEE** page 56

Leaf-curling aphids

■ **PLANTS AT RISK** Apples, damsons, pears, peaches and plums.

■ **RECOGNITION** Immature leaves pucker and curl as aphids feed on plant tissue.

■ **DANGER PERIOD** April to July.

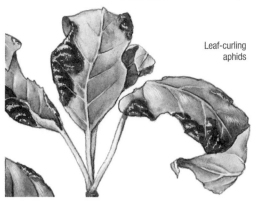

Leaf-curling
aphids

■ **TREATMENT** Spray with an insecticide such as permethrin before the trees blossom and again, if necessary, after blossom appears. **Organic advice** *See* APHIDS (p. 34).

Leaf-rolling rose sawflies

■ **PLANTS AT RISK** All bush and climbing roses.

■ **RECOGNITION** Leaves become tightly rolled up along their length. The sawflies are unlikely to damage the plant.

■ **DANGER PERIOD** May to July.

■ **TREATMENT** Pick off and burn affected leaves as you spot them. Alternatively, spray every fortnight in May with permethrin or liquid derris. **Organic advice** Clear away mulches in autumn and lightly fork over the soil to expose the pupae to predators.

Leaf-rolling
rose sawflies

Lilac leaf miners

■ **PLANTS AT RISK** Lilac (*Syringa*) and privet (*Ligustrum*).

■ **RECOGNITION** The caterpillars of a species of moth burrow into lilac leaves. They also cause leaf-mine blotches on privet leaves.

■ **DANGER PERIOD** June onwards.

■ **TREATMENT** Spray large infestations with imidacloprid. On small outbreaks pick off and burn affected leaves.

Oedema

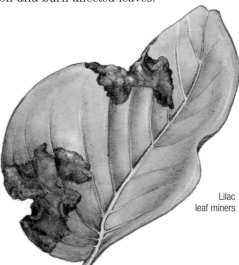

Lilac leaf miners

Oedema (Dropsy)

■ **PLANTS AT RISK** Many species, often succulents, semi-succulents, camellias and ivy-leaved pelargoniums.

■ **RECOGNITION** Small bumps on the leaves are caused by too much water inside the plant. This can be due to a saturated soil or to a wet atmosphere. Leaves sometimes fall off.

■ **DANGER PERIOD** Any time.

■ **TREATMENT** Reduce watering. Increase greenhouse ventilation to reduce humidity. The infected leaves will not recover but they should be left; they will help the plant to lose its excess water.

Onion eelworms

■ **PLANTS AT RISK** Chives, garlic, onions and shallots.

■ **RECOGNITION** Leaves swollen and distorted, bulbs soft and cannot be stored.

■ **DANGER PERIOD** Growing season.

■ **TREATMENT** Pull up and burn affected plants. Rotate crops on a three-year cycle. Grow onions from seeds rather than sets.

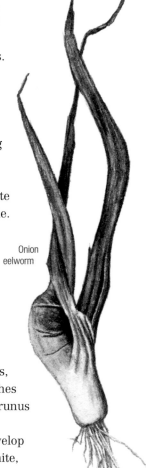

Onion eelworm

Peach leaf curl

■ **PLANTS AT RISK** Almonds, apricots, nectarines, peaches and related ornamental prunus species.

■ **RECOGNITION** Leaves develop large red blisters, turn white, then brown and finally fall prematurely due to fungus.

■ **DANGER PERIOD** Before bud-burst.

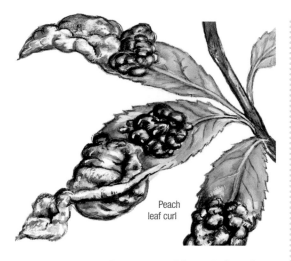

Peach leaf curl

■ **TREATMENT** Remove and burn infected leaves. Spray with mancozeb or Bordeaux mixture in January/early February. Repeat a fortnight later and just before leaf-fall. **Organic advice** Wall-trained trees can be protected from rain with a small roof of polythene. It should be in place before bud-burst, and may be removed once all foliage is expanded. The sides should be left open.

Pear-leaf blister mites

■ **PLANTS AT RISK** Pears, sorbus, cotoneasters and other related trees.
■ **RECOGNITION** Clusters of dark brown pustules appear on the upper surface of the leaves where mites have been burrowing into plant tissue. Young shoots may also be damaged.

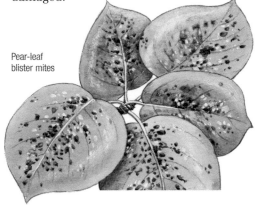

Pear-leaf blister mites

■ **DANGER PERIOD** April to August.
■ **TREATMENT** Pick off and burn infested leaves. If the attack is severe, spray at the end of March with lime-sulphur or with fenitrothion. This is a difficult pest to eliminate, although infested trees may still produce a good crop of pears.

Raspberry virus
▶ **SEE** page 63

Redcurrant blister aphids

■ **PLANTS AT RISK** Red, white and black currants.
■ **RECOGNITION** Irregular red or green blisters appear raised on the surface of leaves. Aphids may be seen underneath leaves, but symptoms will persist after the pests have flown.
■ **DANGER PERIOD** May and June.
■ **TREATMENT** Spray with permethrin or derris. Rake up any fallen leaves and destroy them. **Organic advice** Encourage natural predators of aphids (p. 12).

Redcurrant blister aphids

Stem and bulb eelworms
▶ **SEE** page 119

Smuts

Smuts

■ **PLANTS AT RISK** Herbaceous perennials, bulbs, corms, tubers, onions, sweetcorn.
■ **RECOGNITION** Blister-like swellings form on leaves and stalks, then burst to discharge masses of sooty spores. Caused by a fungal disease. Looks similar to SOOTY MOULD (p. 64).
■ **DANGER PERIOD** Long hot summers.
■ **TREATMENT** Burn diseased plants and plant new ones on a different site. Protect healthy plants by spraying with Bordeaux mixture. Disinfect coldframes or the greenhouse.

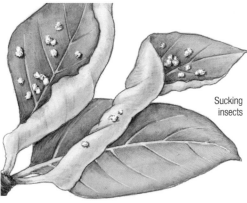

Sucking insects

Sucking insects

▶ **SEE ALSO** BOX SUCKERS (p. 43)
■ **PLANTS AT RISK** Apples and pears.
■ **RECOGNITION** Sap-sucking insects distort leaves and leave sticky excretions. They also infest blossom trusses causing symptoms similar to frost damage.

■ **DANGER PERIOD** April to July.
■ **TREATMENT** Spray fruit trees with permethrin, or derris soon after petal-fall. As an alternative for apple trees, apply a winter wash of tar oil.

Tar spot

Tar spot

■ **PLANTS AT RISK** Acers, especially sycamores.
■ **RECOGNITION** Fruiting bodies of fungi appear as yellow patches on the upper surface of leaves, developing into large black blotches, or red-brown to yellowish blotches.
■ **DANGER PERIOD** When the tree is in leaf.
■ **TREATMENT** Burn diseased leaves as they fall. Spray small trees in spring, when the leaves unfold, with copper-containing fungicide or penconazole or myclobutanil.
Organic advice The effect of the disease is mainly on appearance. No treatment is needed other than for cosmetic reasons.

Tarsonemid mites

■ **PLANTS AT RISK** Begonias, dahlias, fuchsias, gerberas, pot cyclamens, ferns and other greenhouse plants. Outdoors, similar mites attack strawberries and Michaelmas daisies (*Aster*).

Tarsonemid mites

■ **RECOGNITION** Tiny mites feed on the concealed parts of plants causing leaf distortion and discoloration. The mites may also infest the flowers, which then do not open properly.

■ **DANGER PERIOD** Any time of the year under glass.

■ **TREATMENT** There is no effective chemical control, but the symptoms may be relieved with the application of powdered sulphur. Maintain good hygiene in the greenhouse at all times.

Tortrix caterpillars

■ **PLANTS AT RISK** Shrubs, trees and many herbaceous perennials, especially perennial phlox, heleniums and chrysanthemums.

Tortrix caterpillars

Various greenhouse and house plants are also at risk.

■ **RECOGNITION** Leaves and stems are drawn together by silken webs which are spun by 2cm long, grey caterpillars to create protective coverings while they feed.

■ **DANGER PERIOD** May and June outdoors; any time under glass.

■ **TREATMENT** Remove and destroy the caterpillars by hand. Alternatively, spray the plants with fenitrothion or derris.

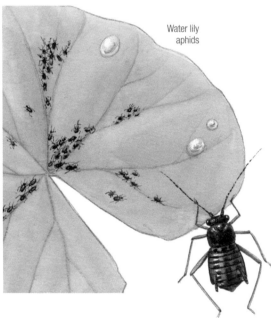

Water lily aphids

Water lily aphids

■ **PLANTS AT RISK** Water lilies (*Nymphea* and *Nuphar*) and pondside plants.

■ **RECOGNITION** Colonies of large, dark-green aphids disfigure the leaves and stems and discolour the flowers.

■ **DANGER PERIOD** Growing season.

■ **TREATMENT** Spray plants frequently with a strong jet of water to dislodge the aphids; fish will eat them. Alternatively, submerge the foliage with sacking or weighted netting for several hours.

Whiptail

- **PLANTS AT RISK** Broccoli and cauliflowers.
- **RECOGNITION** Leaves become distorted – ruffled, thin and strap-like – due to a lack of the mineral molybdenum.
- **DANGER PERIOD** During the growing period in acid soils.
- **TREATMENT** Water with a solution of molybdate – 1 rounded tablespoon in 9 litres of water for every 8m² of soil.

Organic advice As molybdenum is not available to plants growing in acid soils, lime the soil, if necessary, to bring the pH up to 6.5.

White blister

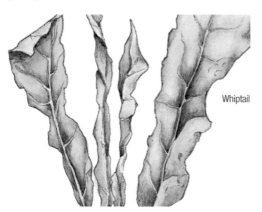

Whiptail

White blister

- **PLANTS AT RISK** Brassicas and related ornamental plants, including honesty (lunaria).
- **RECOGNITION** Leaves and stems develop white glistening pustules filled with the powdery spores of a fungus. Severely affected foliage may be very distorted.
- **DANGER PERIOD** Growing season.
- **TREATMENT** Remove and burn all diseased leaves and stems from the plant.

Adelgids
▶ **SEE** page 71

Anthracnose of cucumbers

■ **PLANTS AT RISK** Cucumbers, melons and marrows grown under glass.
■ **RECOGNITION** Leaves often very pale with transparent spots which become dry and brown in the centre.
■ **DANGER PERIOD** Growing season.
■ **TREATMENT** Burn all infected leaves. Spray or dust with sulphur during the growing season. Reduce greenhouse humidity and provide adequate ventilation. At the end of the season disinfect with cresylic acid or other sterilant.

Anthracnose of cucumbers

Anthracnose of trees
▶ **SEE** page 43

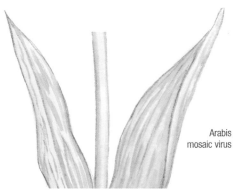

Arabis mosaic virus

Arabis mosaic virus

■ **PLANTS AT RISK** Herbaceous, bulbous and woody plants; fruit crops such as strawberries; vegetables such as celery and lettuce.
■ **RECOGNITION** Mottling of the leaves and plant distortion caused by a virus in the soil and transmitted by a type of eelworm.
■ **DANGER PERIOD** Growing season.
■ **TREATMENT** No chemical control – destroy diseased plants. Avoid growing plants susceptible to infection. Remove plant debris in autumn.

Bacterial canker
▶ **SEE** page 39

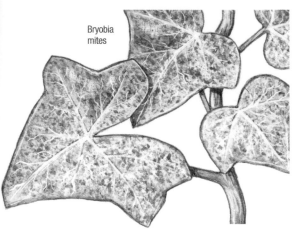

Bryobia mites

Bryobia mites

■ **PLANTS AT RISK** Apples, gooseberries, ivies.
■ **RECOGNITION** The leaves develop a light freckling on the upper surface, and later turn bronze and wither. Unlike red spider mites, bryobia mites feed on the top of the leaf and do not produce silk webbing.
■ **DANGER PERIOD** March onwards throughout the growing season.
■ **TREATMENT** Bryobia mites are difficult to control. Spray with derris or fatty acids during the growing season.

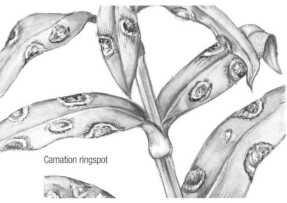

Carnation ringspot

Carnation ringspot

■ **PLANTS AT RISK** Two carnation species are vulnerable to ringspot – *Dianthus barbatus* and *D. caryophyllus*.
■ **RECOGNITION** Circular grey spots develop on the foliage, along the stems and occasionally on the flowers. Foliage may shrivel under damp conditions. Grey fungus may also appear in the centre of the spots.
■ **DANGER PERIOD** Growing season.
■ **TREATMENT** Remove diseased leaves and spray regularly with Bordeaux mixture, following the instructions on the label.

Carrot flies
▶ **SEE** page 113

Celery leaf miners

Celery leaf miners (Celeryflies)

■ **PLANTS AT RISK** Celery, carrots and parsnips.
■ **RECOGNITION** Brown blotches on foliage caused by small white maggots tunnelling through the leaves, which shrivel and die. Severe outbreaks can destroy plants.
■ **DANGER PERIOD** From May to autumn.
■ **TREATMENT** Spray plants with permethrin or bifenthrin when symptoms are first identified, or cut off and burn affected leaflets. **Organic advice** Pick off infected leaves, or squash the larvae within them. If the pest is a regular problem, grow celery crops under fine netting or horticultural fleece.

Celery leaf spot

Mineral chlorosis

Celery leaf spot

■ **PLANTS AT RISK** Celery and celeriac – mature plants and seedlings.
■ **RECOGNITION** Small brown spots on leaves and stems develop into black fruiting bodies of fungus.
■ **DANGER PERIOD** Wet weather.
■ **TREATMENT** Spray seedlings and plants with Bordeaux mixture or mancozeb, repeating at weekly intervals if necessary. Feed established plants with high-potash fertiliser to avoid soft growth. **Organic advice** Clear up all crop debris. Do not overuse nitrogen-rich feeds. Destroy infected plants at the end of the season.

Chlorosis (Mineral)

■ **PLANTS AT RISK** Hydrangeas, peaches, ceanothus, raspberries, acid-loving plants – such as camellias and rhododendrons – on alkaline soil and many others.
■ **RECOGNITION** Leaves lose their rich green colour to become pale yellow or white.
■ **DANGER PERIOD** Growing season.
■ **TREATMENT** Dig peat, pulverised bark or crushed bracken into the soil and use only fertilisers sold as lime-free. For iron deficiency (the most common form of chlorosis) use a chelated-iron compound.

Chlorosis (Viral)

■ **PLANTS AT RISK** Vast range of garden plants.
■ **RECOGNITION** The foliage yellows, either across the whole leaf, on the margins or just on the veins. The yellowing may occur in patterns of lines or rings.

Viral chlorosis

■ **DANGER PERIOD** Growing season.
■ **TREATMENT** Control aphids (see p. 34) which spread the infection. Destroy badly infected plants, and buy virus-free plants where they exist.

Club root

► **SEE** page 88

Cox spot

■ **PLANTS AT RISK** 'Cox's Orange Pippins' and (less so) some other apple varieties.
■ **RECOGNITION** Small, round tan-coloured spots on leaves due to an unknown physiological disorder.
■ **DANGER PERIOD** Usually after drought.
■ **TREATMENT** Try watering during dry spells and mulching the trees.

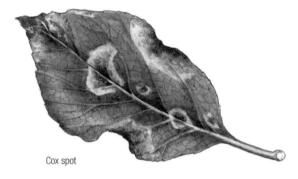

Cox spot

Crinkle

■ **PLANTS AT RISK** Strawberries.
■ **RECOGNITION** Yellow spots appear on the leaves, the centre of which may become red or purple. When these spots multiply and turn brown and the leaves pucker, the outbreak is severe. The cause is an aphid-borne virus which stunts growth.
■ **DANGER PERIOD** Growing season.
■ **TREATMENT** Burn diseased plants and protect remainder by controlling aphids.

Crinkle

Crown rot of rhubarb

▶ **SEE** page 88

Dianthus leaf spot

■ **PLANTS AT RISK** *Dianthus barbatus*, *D. caryophyllus* and occasionally *D. chinensis*.
■ **RECOGNITION** Leaves develop circular or oval spots with tiny black spores, surrounded by a purple border, or small, round purple spots that gradually increase and coalesce. Infected leaves wither and die off at the tips. Caused by several fungi.
■ **DANGER PERIOD** During damp periods of the growing season.
■ **TREATMENT** Remove infected leaves and spray plants with Bordeaux mixture or mancozeb.

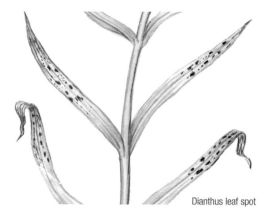

Dianthus leaf spot

Downy mildew

■ **PLANTS AT RISK** Brassica seedlings, marrows, courgettes and onions.
■ **RECOGNITION** Plants develop a grey or whitish furry covering on the underside of the leaves, with blotches of yellow on the upper side, the result of being overcrowded and inadequately ventilated. Onion bulbs may rot in store.
■ **DANGER PERIOD** During growing season, and on onions in store.

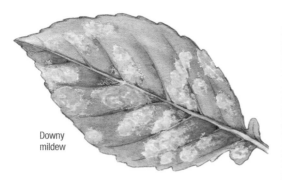
Downy mildew

■ **TREATMENT** Thin out plants and spray with carbendazim or mancozeb (wait 14 days before eating). **Organic advice** Rotate future crops. Grow resistant varieties where available. Always sow seeds in fresh sterilised compost, and improve ventilation when raising plants under glass. Ensure plants are adequately spaced.

Eelworms
▶ **SEE** page 45

Fire blight

■ **PLANTS AT RISK** Cotoneasters, hawthorns, sorbus species, apples, pears and other related ornamentals.
■ **RECOGNITION** Flowers turn black, leaves turn brown and wither. Golden or white slime may exude from the stems. The symptoms can appear suddenly, giving the plant the appearance of having been scorched by fire.

Fire blight

■ **DANGER PERIOD** Flowering time.
■ **TREATMENT** Remove and destroy the diseased shoots when the symptoms are identified, cutting back beyond the damage by at least 30cm. If the whole plant shows symptoms, dig it up and remove it. Plant a resistant variety, such as *Sorbus intermedia* or *Pyrus calleryana* 'Bradford'.

Fruit-tree red spider mites

Fruit-tree red spider mites

■ **PLANTS AT RISK** Apples, plums, pears, damsons, cotoneasters, hawthorns, sorbus.
■ **RECOGNITION** Leaves turn bronze and wither. In a bad attack, they fall off in July. The mites are invisible to the naked eye. Tiny red spiders that run over leaves are called brick spiders and do no harm.
■ **DANGER PERIOD** April to September.
■ **TREATMENT** Spray immediately after flowering with permethrin, derris or a fatty acid spray. **Organic advice** Mulch trees, and water in dry weather. Natural predators in the garden will usually keep the pest under control.

Frost damage
▶ **SEE** page 74

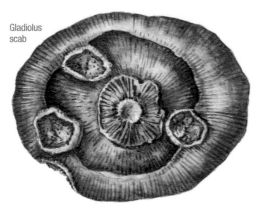

Gladiolus scab

Gladiolus scab

■ **PLANTS AT RISK** Gladioli.

■ **RECOGNITION** Reddish specks appear on the leaves, which enlarge and darken. Leaf tips shrivel. Craters develop at the base of the corms with raised rims and glossy coating. Yellow spots on the corms exude gum.

■ **DANGER PERIOD** Scab strikes in summer but may not reveal itself until corms are lifted and stored.

■ **TREATMENT** As for DRY ROT (p. 114).

Gladiolus yellows

■ **PLANTS AT RISK** Gladioli.

■ **RECOGNITION** Yellow stripes appear on the leaves, which gradually turn completely yellow and then die back. The flower stems may be crooked and greener than normal.

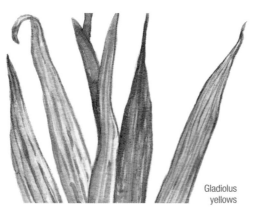

Gladiolus yellows

■ **DANGER PERIOD** During the growing season, but the disease develops in storage.

■ **TREATMENT** Burn the infected plants immediately. Grow gladioli corms on a fresh site each year, and avoid nitrogen-rich fertilisers and manures.

Gooseberry mildew (European)

▶ **SEE** page 100

Grey mould (Botrytis)

Grey mould (Botrytis)

■ **PLANTS AT RISK** Many types, both outdoors and under glass.

■ **RECOGNITION** A grey velvety mould, caused by a fungal disease, forms on leaves, flowers and soft fruits. The disease is encouraged by damp, overcrowded conditions and poor ventilation in a greenhouse.

■ **DANGER PERIOD** Growing season. Worst in winter for lettuces.

■ **TREATMENT** If possible destroy infected plants. Pick off dying flowers and buds. Spray herbaceous plants with carbendazim. Destroy diseased bulbs; dust healthy ones with mancozeb. Cut out infected areas on trees or shrubs. Spray soft fruit with mancozeb when flowers appear, and until just before fruits ripen. Under glass, spray with mancozeb as symptoms appear.

Hard rot
▶ SEE page 115

Holly leaf miners
■ **PLANTS AT RISK** Holly.
■ **RECOGNITION** White blotches on the leaves caused by maggots tunnelling through them. The leaves may fall from young trees.
■ **DANGER PERIOD** Late May to July.
■ **TREATMENT** Spray regularly with imidacloprid, following manufacturer's instructions.

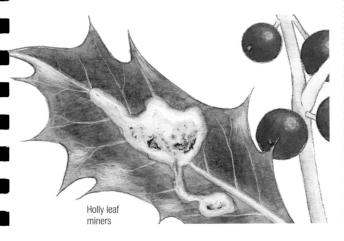

Holly leaf miners

Ink disease
▶ SEE page 91

Iris mosaic virus
■ **PLANTS AT RISK** Bulbous and bearded irises.
■ **RECOGNITION** Yellow streaks and spots occur on young leaves. Plants may be stunted. The flowers often show darker streaks on the normal ground colour.
■ **DANGER PERIOD** Growing season.
■ **TREATMENT** Destroy all infected plants. Control aphids which spread the virus (see p. 34).

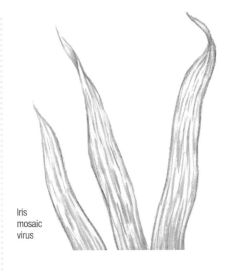

Iris mosaic virus

Leaf eelworms
■ **PLANTS AT RISK** Particularly troublesome in greenhouses with begonias, coleus, gloxinias, ferns and other foliage plants. Outdoors: mint, strawberries and other soft fruit.
■ **RECOGNITION** Brown or black blotches between leaf veins, and distortion of buds and young growth, caused by tiny pests.
■ **DANGER PERIOD** Winter.
■ **TREATMENT** No cure; severely infested plants should be burnt. Maintain good hygiene and grow plants in dry conditions.

Leaf eelworms

Leaf mould
▶ SEE page 108

Leaf rot

Leaf rot

■ **PLANTS AT RISK** Dianthus species, including garden pinks.

■ **RECOGNITION** Leaves and stems develop white or brownish water-soaked spots with dark margins, or grey spots. Flower buds may also be affected.

■ **DANGER PERIOD** Growing season.

■ **TREATMENT** Remove and burn all diseased tissues and spray plant with Bordeaux mixture.

Leaf spot

■ **PLANTS AT RISK** Blackcurrants and gooseberries, celery and spinach, plus a wide range of ornamental plants.

Leaf spot

■ **RECOGNITION** The leaves and stems develop round or oval brown spots – sometimes with a black pinpoint – and the leaves then fall prematurely.

■ **DANGER PERIOD** Late spring onwards.

■ **TREATMENT** Remove and burn all diseased leaves. Spray with mancozeb, penconazole or Bordeaux mixture, following the instructions on the label.

Leafhoppers

Leafhoppers

▶ **SEE ALSO** RHODODENDRON LEAFHOPPERS (p. 38)

■ **PLANTS AT RISK** Roses, pelargoniums, primulas and others, both under glass and in the open.

■ **RECOGNITION** Coarse white flecks on leaves and – often on the underside – the cast-off skins of the insects.

■ **DANGER PERIOD** April to October, but any time under glass.

■ **TREATMENT** Spray with permethrin, repeating if necessary at fortnightly intervals or use a systemic insecticide, such as imidacloprid.

Roses

SOME ROSES ARE PRONE TO MORE THAN THEIR FAIR SHARE OF AILMENTS, ESPECIALLY WHERE THEY ARE GROWING IN LESS THAN IDEAL SITUATIONS. A FEW SIMPLE PRECAUTIONS CAN STOP A MINOR OUTBREAK DEVELOPING INTO A SERIOUS PROBLEM.

The red-tinged early shoots on rose bushes and climbers are a sign of healthy new growth but also a magnet for greenfly, so keep an eye open.

Choose disease-resistant varieties, some of which have remarkably good health under normal garden conditions.

Buy healthy plants and plant them in well-prepared sites in the garden.

Inspect plants regularly, in particular shrub roses, which often begin growing first and offer an early bridgehead for aphids and diseases.

Feed plants at the right time; avoid over-feeding or the excessive use of high-nitrogen fertilisers, which can cause roses to develop soft, vulnerable growth.

Clear up all leaves, dead material and prunings promptly, especially where diseases like black spot have occurred before.

Treat problems straight away, choosing the right product and measuring out and applying it exactly according to the instructions.

If you grow a lot of roses, consider starting a regular spraying plan early in the season to prevent outbreaks.

Disperse roses in a mixed border to reduce the incidence of diseases.

Many roses are prone to particular ailments. It is important to recognise the symptoms so that you can deal with them promptly, but, better still, try to prevent them taking hold (see above).
1 black spot
2 mildew **3** greenfly

Take preventative measures by spraying new growth against common rose problems.

DISEASE-RESISTANT VARIETIES

- 'Albéric Barbier'
- Bonica
- 'Buff Beauty'
- 'Cécile Brünner'
- 'Charles de Mills'
- 'Félicité Perpétue'
- 'Flower Carpet' series
- 'Fritz Nobis'
- Gertrude Jekyll
- Just Joey
- 'Maigold'
- 'Paul's Himalayan Musk'
- 'Penelope'
- Remembrance
- 'Roseraie de l'Haÿ'
- 'Tickled Pink'
- Winchester Cathedral

Lilac blight

Lilac blight

- **PLANTS AT RISK** Syringas.
- **RECOGNITION** Brown angular spots develop on the leaves, followed by young shoots turning black and withering.
- **DANGER PERIOD** Spring.
- **TREATMENT** Cut the infected shoots back to healthy buds and spray with Bordeaux mixture. Spray again the following spring when the leaves start to appear.

Lily disease

- **PLANTS AT RISK** Lilies – particularly *Lilium candidum*.
- **RECOGNITION** Leaves develop small red-brown oval-shaped, water-saturated spots of fungus which spread rapidly. In humid conditions, leaves may turn brown, flowers rot and stems collapse.
- **DANGER PERIOD** Wet seasons.
- **TREATMENT** Cut off and burn infected growth. When new leaves appear in spring, spray all plants with mancozeb or Bordeaux mixture. Repeat at fortnightly

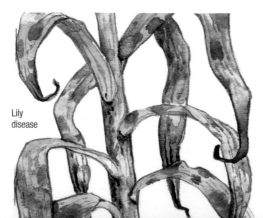

Lily disease

intervals until flowering begins and, during a wet season, again after flowering. Destroy all infected material at the end of the season, as the fungus can overwinter in the soil.

Magnesium deficiency

- **PLANTS AT RISK** All types of plants, particularly tomatoes and apples.
- **RECOGNITION** Leaves turn yellow (sometimes orange-brown) between the veins, giving a marbled effect; they then fall prematurely. Symptoms appear first on older leaves. In light, acid, sandy soil the problem is worse in a wet season as magnesium is easily leached.
- **DANGER PERIOD** During the growing season, or following the use of a high-potash fertiliser.
- **TREATMENT** Spray the plants with a solution of magnesium sulphate – 8 tablespoons to 11 litres of water – plus a few drops of detergent. **Organic advice** If lime is needed, use dolomitic (magnesian) limestone.

Magnesium deficiency

Manganese deficiency

- **PLANTS AT RISK** Many species.
- **RECOGNITION** Yellowing occurs between the veins of older leaves. Dead patches may appear among the yellow areas. The

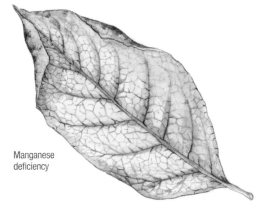

Manganese
deficiency

disorder occurs mainly in poorly drained sandy soil, highly organic soil and wet areas of low acidity.

■ **DANGER PERIOD** Growing season.

■ **TREATMENT** Spray the plants with a solution of manganese sulphate – 2 tablespoons to 11 litres of water plus a few drops of detergent – or apply chelated or fritted compounds.

Nitrogen deficiency

■ **PLANTS AT RISK** All types, but most commonly fruit trees and vegetables.

■ **RECOGNITION** The young leaves turn pale yellow-green, and later develop yellow, red or purple tints. The plants are stunted and lack vigour.

■ **DANGER PERIOD** Growing season.

■ **TREATMENT** Use nitrogenous fertiliser such as blood, fish and bone or sulphate of ammonia in spring. Improve soil structure and fertility generally. For a quick result, water with a liquid feed or apply a nitrogen-rich foliar feed. A temporary deficiency can arise in cold weather, but it will disappear as the soil warms up.

Nitrogen
deficiency

Physiological disorders

HEALTHY PLANTS depend on a combination of water, mineral salts, air, temperature and light. If any of these essentials is lacking or the balance is wrong, the result may produce a physiological disorder.

Correct watering must be available at each stage of a plant's growth. Too much, too little or sporadic watering can create problems from which a plant may never recover.

Mineral salts should be present in the soil in the correct quantities and form. Some plants need more or less nutrients than others, according to the nature of the soil, the growing conditions and the care they are given. Discoloured foliage and poor growth often indicate some kind of mineral deficiency.

An atmosphere that is too humid can lead to the growth of fungal diseases, while one that is too dry may cause poor growth and induce flower buds to drop. Fluctuating temperatures that are either too high or too low can cause similar problems, while a lack of light leads to thin, weak and colourless plants, with the flowering reduced or non-existent.

Physiological disorders can usually be rectified by careful cultivation – feeding, mulching, watering, drainage and correct spacing.

Dry, discoloured foliage is often indicative of a physiological disorder such as lack of water or a mineral deficiency.

Pear scab
▶ **SEE** page 102

Phytophthora root rot
▶ **SEE** page 92

Potassium deficiency

■ **PLANTS AT RISK** Plants in clay, chalky soil or peat, particularly those that need extra potash – potatoes, beans, tomatoes, apples, pears, currants.

■ **RECOGNITION** Stunted growth and reduced leaf size. The leaves turn a dull blue-green, and turn brown at the tips or margins, or curl downwards. Shoots, and sometimes whole branches, die back. Fruit are sparse, small, highly coloured.

■ **DANGER PERIOD** Any time.

■ **TREATMENT** For fruit, apply sulphate of potash every March at 100–140 g per m². To improve flowering or berrying on trees and shrubs, apply sulphate of potash at 18 g per m² in early spring or late summer.

Potassium deficiency

Powdery mildew

■ **PLANTS AT RISK** A wide range of plants including strawberries, roses and apples.

Powdery mildew

■ **RECOGNITION** A white mildew, caused by a fungus, appears on leaves, shoots and flowers, and weakens the plant. Warm dry weather encourages the disease.

■ **DANGER PERIOD** Growing season.

■ **TREATMENT** Remove and burn infected growth. Water well and mulch with garden compost, leaf-mould or well-rotted manure. Avoid overuse of nitrogenous fertiliser. Spray with mancozeb, following the instructions on the label. Sulphur can be applied as a dust. Buy resistant varieties of rose, apple and strawberry plants.

Pyracantha scab

■ **PLANTS AT RISK** Pyracanthas.

■ **RECOGNITION** Leaves and berries acquire a thick, felt-like, olive-brown or black coating of fungus and scabby lesions form on the shoots. Leaves may drop early and berries become disfigured and shrivelled.

Pyracantha scab

- **DANGER PERIOD** During wet periods.
- **TREATMENT** Cut out and burn all diseased shoots. Spray with myclobutanil once a fortnight as soon as the disease is seen.
Organic advice Don't plant pyracanthas in damp places. Grow resistant varieties.

Raspberry virus

- **PLANTS AT RISK** Raspberries.
- **RECOGNITION** Leaves develop yellow blotches and become distorted. Canes are stunted and the crop is poor.
- **DANGER PERIOD** Growing season.
- **TREATMENT** Dig up and burn all affected plants. Plant new canes on a fresh site and control aphids, which spread the virus. Only buy raspberry canes which have been certified free from virus.

Raspberry virus

Rhododendron lacebugs

- **PLANTS AT RISK** Young and adult rhododendrons.
- **RECOGNITION** Fine mottling on the upper surface of leaves, with a rusty-brown or chocolate spotting on the underside.
- **DANGER PERIOD** May and June.
Cut out and burn infested branches in May. Spray in late May with permethin, and again in mid June.

Rhododendron lacebugs

Organic advice Don't grow rhododendrons in exposed, dry or sunny sites. Some species and hybrids are less vulnerable than others; grow them if the pest is known to be a problem.

Rose black spot

- **PLANTS AT RISK** Roses.
- **RECOGNITION** Leaves develop sooty irregular spots on both surfaces which, when severe, can lead to yellowing and leaf-fall. Caused by a fungal disease.
- **DANGER PERIOD** Growing season.
- **TREATMENT** Spray with penconazole, myclobutanil or mancozeb in February and then at two-weekly intervals. **Organic advice** Grow resistant varieties. Do not overfeed with nitrogen-rich fertilisers. Remove dead leaves from the soil in autumn. In spring, cut out infected shoots before bud-burst, remove old leaves and apply mulch.

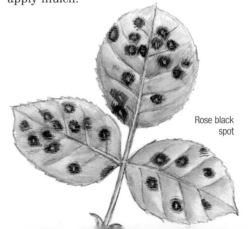

Rose black spot

Rust

▶ **SEE** page 69

Salix watermark

▶ **SEE** page 93

Scorching (Scorch)

■ **PLANTS AT RISK** Most types of greenhouse and indoor plants, plus acers and beeches.
■ **RECOGNITION** Affected leaves develop pale brown spots or become papery, either due to cold drying winds or – often when in a greenhouse – scorching by the sun.

Scorching

■ **DANGER PERIOD** Spring for trees and shrubs; summer for greenhouse and indoor plants.
■ **TREATMENT** Shade the greenhouse or remove vulnerable plants. Ensure that plants are adequately watered, but avoid wetting the leaves during bright warm periods. Water indoor plants from the base to avoid wetting the leaves.

Silver leaf

■ **PLANTS AT RISK** Peaches, plums, cherries and other prunus species; apples, pears and lilacs; and other trees and shrubs.
■ **RECOGNITION** Some leaves become silvered and later turn brown, while infected branches die back. A flat purple fungus develops on dead wood.
■ **DANGER PERIOD** September to May.

Silver leaf

■ **TREATMENT** Cut a branch, at least 2.5cm thick, and moisten the wound. If it is diseased a brown or purple stain will appear. Cut out the affected branches to a point 15cm below the fungus. Use tri-sodium orthophosphate to sterilise tools. Prune plums and cherries only in June–August. Try biological control with trichoderma pellets.

Sooty mould

■ **PLANTS AT RISK** Very wide range of plants, both outdoors and under glass.
■ **RECOGNITION** Dark brown or black sooty fungus appears on the top side of the leaves. It is always associated with sap-feeding insects such as aphids, scale insects, whiteflies and mealy bugs.
■ **DANGER PERIOD** Throughout the year.
■ **TREATMENT** Wipe off the mould with a damp cloth. Apply a systemic insecticide, such as imidacloprid, to control the pest that is feeding on the plant.

Sooty mould

Tar spot
▶ **SEE** page 48

Tarsonemid mites
▶ **SEE** page 48

Thrips
▶ **SEE** page 84

Tree rust

■ **PLANTS AT RISK** Mainly birch, plum, poplar and willow.

■ **RECOGNITION** Masses of brown, orange or yellow spores, caused by fungi on the underside of leaves and stems. Serious outbreaks can kill plants.

Tree rust

■ **DANGER PERIOD** Summer.

■ **TREATMENT** Destroy severely infected plants. Remove affected leaves on mild outbreaks and spray with penconazole or mancozeb when symptoms are first seen. Encourage healthy, vigorous growth with good cultural treatment.

Violet root rot
▶ **SEE** page 95

White rust

White rust

■ **PLANTS AT RISK** Dendranthemas (chrysanthemums).

■ **RECOGNITION** The upper leaf surface develops yellow to pale green spots. The underside of the leaf has buff or white spots. The spots may turn brown and die in the centre.

■ **DANGER PERIOD** Growing season.

■ **TREATMENT** Cut off the infected parts of the plant and destroy them. Spray the remainder of the plant with a fungicide such as myclobutanil to kill off any incipient infection.

Cutworms

▶ **SEE** page 74

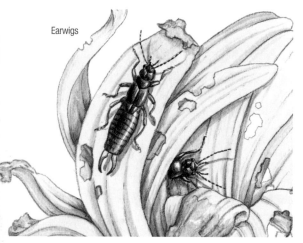

Earwigs

Earwigs

■ **PLANTS AT RISK** Dendranthemas (chrysanthemums), clematis, dahlias and other ornamentals.

■ **RECOGNITION** Petals and leaves chewed and left in tatters by brown, beetle-like insects, about 1cm long.

■ **DANGER PERIOD** May to October.

■ **TREATMENT** Spray or dust plants with derris or bifenthrin. Trap earwigs in rolls of corrugated cardboard, old sacking or flowerpots filled with straw, and then destroy them. **Organic advice** Control earwigs only if they are causing harm. They eat greenfly and are valuable pest controllers in their own right.

Lily beetles

▶ **SEE** page 36

Slugs and snails

■ **PLANTS AT RISK** Many garden plants. Tender shoots particularly vulnerable, but also roots, tubers, bulbs and corms.

■ **RECOGNITION** Irregularly shaped holes eaten in bulbs, roots, stems, leaves and flowers. Slime trails reveal their active presence. Slugs and snails usually feed at night, hiding by day.

■ **DANGER PERIOD** Mild wet weather in spring and autumn.

■ **TREATMENT** Remove decaying plant material and scatter metaldehyde pellets, or a slug killer based on aluminium sulphate. **Organic advice** Avoid heavy dressings of manure or mulch around young plants. Never transplant seedlings into cold wet soil. Protect young plants with copper rings or cloches made from plastic water bottles. Encourage predators (p. 12). See right for more control tips.

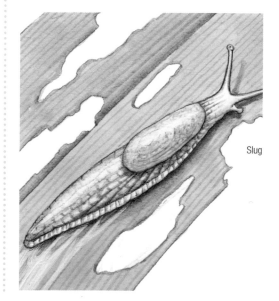

Slug

Slugs and snails

SHELTERING IN THICK FOLIAGE DURING THE DAY, SLUGS AND SNAILS SLIDE INTO ACTION AT NIGHT TO DEVOUR DELPHINIUMS, HOSTAS, LUPINS, TENDER EMERGING SEEDLINGS AND CABBAGE HEARTS.

SINCE EACH SLUG OR SNAIL produces 500 offspring every season, it is imperative to control the population in your garden from early spring.

One of the most effective ways of dealing with these pests is to go out on warm, damp summer evenings when they are most active and collect them by hand with the help of a torch. Put them into a steel bucket or tin and pour in boiling water.

Slugs and snails will not cross a rough-surfaced barrier, such as one made from gravel, crushed eggshells, nut shells, sharp sand, cinders, lime or crushed oyster shells. Spread any of these materials round vulnerable plants in the garden.

Another approach is to trap the slugs and snails before they reach vulnerable plants. They find beer irresistible, so bury a plastic cup, half-filled with beer in the soil close to the plants that need to be protected. To prevent beneficial insects such as ground beetles falling in, cover the cup with an upturned plastic flower pot with a large hole in the bottom. Empty the cup and renew the beer every three days.

Empty grapefruit halves can also be used as traps. Place upturned halves near threatened plants. Attracted by the smell and moisture, slugs and snails will congregate inside the grapefruit halves, ready for collection and disposal in the morning.

Hedgehogs, frogs and toads eat slugs and snails, so create homes for them by leaving a patch of wild garden or by installing a small pond. Attract snail-eating birds, such as thrushes, with berry-bearing trees and shrubs to nest in and to provide food and shelter.

Slugs are particularly attracted to newly emerging delphiniums, hostas and peonies. You may wish to use a liquid slug-killer on these plants. Spray the plants and soil when the growing shoots are 2.5cm above ground level and repeat the application when the shoots are 7.5cm high. If you are an organic gardener, apply an aluminium sulphate slug killer or a biological control agent (p. 37).

Modern slug pellets are also effective, but have the potential to harm other wildlife. They are usually coloured blue to discourage animals from eating them and also contain a chemical deterrent. The pellets are best used sparingly in the garden, distributed about 10–15cm (4–6in) apart.

If you cannot find it in your heart to kill slugs and snails, but still need to get rid of them, simply collect the pests and dump them on a piece of waste ground.

1 Scatter grit around young plants, using a pot as a guide.
2 Buy a proprietary slug pot filled with a toxic mixture.
3 Cut a serrated collar from a plastic bottle using pinking shears.
4 Place the base of a clear plastic bottle filled with old beer near young plants, to act as a trap – or use upturned grapefruit skins.

Carnation flies
(Carnation leaf miner)

■ **PLANTS AT RISK** Carnations and pinks.
■ **RECOGNITION** Maggots tunnel into the leaves and stems causing infected shoots to wilt and die.
■ **DANGER PERIOD** Throughout the year.
■ **TREATMENT** Carnation flies are not easily eradicated. Spray with imidacloprid as soon as the miners appear on the foliage in September.

Carnation flies

Chafer beetles
▶ **SEE** page 87

Chrysanthemum stool miners

■ **PLANTS AT RISK** Dendranthemas (chrysanthemums).
■ **RECOGNITION** Growth is inhibited by light yellow maggots that tunnel into the stools of the plant and cause damage to the roots.

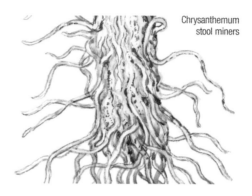

Chrysanthemum stool miners

■ **DANGER PERIOD** September to May.
■ **TREATMENT** Spray. After lifting stools to take cuttings, water them with a spray-strength solution of imidacloprid.

Fasciation

■ **PLANTS AT RISK** Many, especially delphiniums, forsythias, lilies, prunus species and roses.
■ **RECOGNITION** Stems flat and distorted.
■ **DANGER PERIOD** Growing season.
■ **TREATMENT** Treatment is not necessary. Affected areas of shrub stems may be cut back to improve their appearance.

Fasciation

Gooseberry mildew (American)
▶ **SEE** page 99

Leafy gall
▶ **SEE** page 79

Leatherjackets
▶ **SEE** page 91

Leek rust

Leek rust
■ **PLANTS AT RISK** Leeks.
■ **RECOGNITION** Shoots become twisted or malformed and covered with an orange powder.
■ **DANGER PERIOD** Summer and early autumn, or in spring on overwintered crops.
■ **TREATMENT** If rust attacks late varieties of leeks, spray with mancozeb. **Organic advice** Symptoms often disappear in autumn, so no treatment is necessary. Leek rust is more likely where potash is low, nitrogen is high and drainage is poor.

Lilac blight
▶ **SEE** page 60

Mint rust
■ **PLANTS AT RISK** Mint.
■ **RECOGNITION** Thickened and distorted shoots bear orange-coloured fungal spores.
■ **DANGER PERIOD** Symptoms appear in

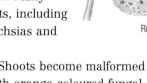

Mint rust

spring; affected plants are permanently diseased.
■ **TREATMENT** Cut out and burn affected shoots, and spray the soil with mancozeb. Burn off withered top growth of the mint bed in autumn or early winter.

Pear scab
▶ **SEE** page 102

Pear-leaf blister mites
▶ **SEE** page 47

Root aphids
▶ **SEE** page 111

Rust
▶ **SEE ALSO**
LEEK RUST,
MINT RUST,
WHITE RUST
(p. 65)
■ **PLANTS AT RISK** Many decorative plants, including rose bushes, fuchsias and hollyhocks.

Rust

■ **RECOGNITION** Shoots become malformed and covered with orange-coloured fungal spores.
■ **DANGER PERIOD** Growing season.
■ **TREATMENT** Remove all the affected stems. Spray plants with penconazole or mancozeb at the first sign of an attack, following the instructions on the label.

Sucking insects
▶ **SEE** page 48

Powdery mildew

▶ **SEE** page 62

Woolly aphids

Woolly aphids

■ **PLANTS AT RISK** Apple trees and related ornamental trees and shrubs, including crab apples, cotoneasters and pyracanthas.

■ **RECOGNITION** Tufts of waxy white wool on trunks, branches and twigs produced by aphids. Galls can also occur and may split, allowing diseases to enter the plant.

■ **DANGER PERIOD** April to September.

■ **TREATMENT** Immediately wool appears, spray with a high volume of bifenthrin or permethrin. A difficult pest to control.

Organic advice Encourage natural aphid predators (p. 12). Cut out severely galled branches. Spray with insecticidal soap under high pressure.

Ladybirds EQUALLY AT HOME IN MEADOWS, WOODS AND CITIES, ladybirds are especially welcome in the garden, where they and their larvae are voracious devourers of aphids. They are found in large numbers, partly due to the fact that they have few natural enemies – their striking coloration serves as a warning to birds that they are evil-tasting and poisonous.

A female ladybird lays about 200 eggs on aphid-infested plants. Her slate-blue larvae eat hundreds of aphids over their three-week lives. Providing your garden is not overrun with aphids, ladybirds can ensure an efficient and organic way of controlling them.

There are some 40-odd species of ladybird in Britain, plus a few more that migrate annually from the Continent. They are mostly distinguished by their coloration and the number of spots on their wing-cases. There is a vegetarian 24-spotter, for example, and a totally black race of ladybirds that live in Glasgow and Merseyside, whose hue is claimed by the locals to be due to lack of sun. All are very much the friends of gardeners and should be encouraged (see p. 12).

Don't be too tidy in the garden. Leave some areas of fallen leaves or spent herbaceous perennials where ladybirds can shelter over the winter.

Adelgids

- **PLANTS AT RISK** Many conifers, such as larch, pine, spruce and fir.
- **RECOGNITION** Small dark insects, related to aphids, suck the sap of the leaves and stems. They excrete a sticky honeydew on which sooty moulds develop. In early summer, colonies cluster beneath white waxy wool.
- **DANGER PERIOD** April and May.
- **TREATMENT** Spray thoroughly once with bifenthrin from early March through to May, and again about three weeks later.

Apricot die-back

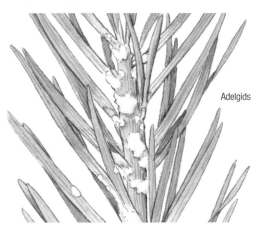

Adelgids

Apricot die-back

- **PLANTS AT RISK** Apricots.
- **RECOGNITION** Branches exude large amounts of gum, and die. The condition is often caused by fungi entering the branches through a wound.
- **DANGER PERIOD** Throughout the year.
- **TREATMENT** Cut out and burn all the dead wood. Avoid pruning trees in winter. If

appropriate, improve cultural conditions, because the disease can arise if the plant is suffering from malnutrition.

Bark splitting

- **PLANTS AT RISK** Many species, including fruit trees.
- **RECOGNITION** The tree's bark splits and fissures open up.
- **DANGER PERIOD** Any time.
- **TREATMENT** Cut out dead wood and remove loose bark to reveal a clean wound. Feed, mulch and water the tree properly and the wound should heal naturally.

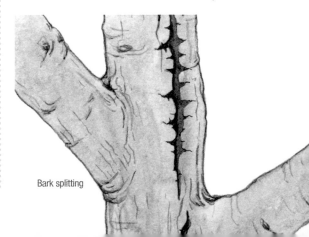

Bark splitting

Beech bark disease

■ **PLANTS AT RISK** Young and mature beech trees.

■ **RECOGNITION** Insect feeding holes in the bark create wounds through which fungus enters. Wounds weep at first, followed by yellowing of foliage and red-black bumps on the bark. Branches die back and the tree eventually dies.

■ **DANGER PERIOD** Late spring.

■ **TREATMENT** In the early stages, insect attacks can be prevented by spraying with permethrin or bifenthrin during the growing season.

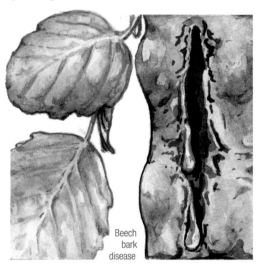

Beech bark disease

Birch polypore

■ **PLANTS AT RISK** Birch trees.

■ **RECOGNITION** Die-back caused by grey-brown horseshoe-shaped bracket fungus, up to 15cm across. The fungus attacks dead wood and spreads decay.

■ **DANGER PERIOD** Any time.

■ **TREATMENT** Cut out and destroy dead wood. Mulch trees regularly with well-decayed manure.

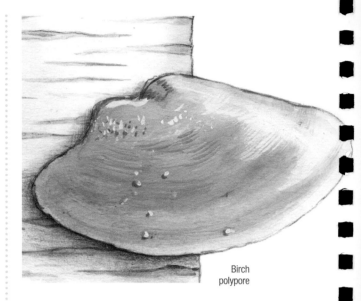

Birch polypore

Bracket fungus

■ **PLANTS AT RISK** All trees, especially alder, ash, beech, oak, poplar and – particularly vulnerable – birch.

■ **RECOGNITION** Fungi, up to 30cm across, sprout from trunks and branches of trees. The spores spread through dead or damaged tissues and may cause die-back.

■ **DANGER PERIOD** July to October, but fungi may appear months after infection.

■ **TREATMENT** Bracket fungus can be a symptom of many different diseases. Seek professional advice.

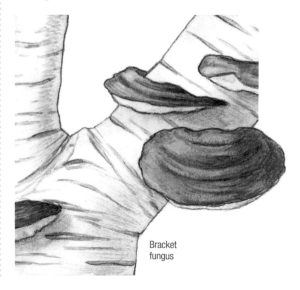

Bracket fungus

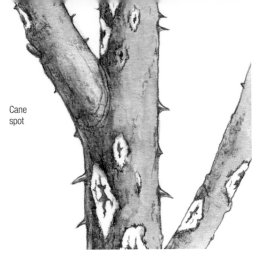
Cane spot

Cane spot

■ **PLANTS AT RISK** Raspberries, loganberries and hybrid berries.
■ **RECOGNITION** A bacterial infection causes round purple spots, which later form cankers on canes; spots with a whitish centre on leaves; and misshapen fruits.
■ **DANGER PERIOD** May to October.
■ **TREATMENT** Cut out and burn badly infected canes. Spray with copper spray fortnightly from bud-burst to petal-fall.

Canker

▶ **SEE ALSO** BACTERIAL CANKER (p. 39), PARSNIP CANKER (p. 117), STEM CANKER (p. 76)
■ **PLANTS AT RISK** Many trees and shrubs, including apple, ash and poplar.
■ **RECOGNITION** Brown, cracked and sunken patches appear on the bark. Slime oozes from the area. Encirclement of a shoot or trunk causes die-back. Bleeding canker of horse chestnut is a spreading problem – young trees are most vulnerable.

Canker

■ **DANGER PERIOD** Any time.
■ **TREATMENT** Cut out and burn infected areas and small branches. Spray with Bordeaux mixture or copper fungicide, following the instructions on the label. If a tree is badly affected, destroy it. **Organic advice** Do not grow apple trees on wet, badly drained soil. *Populus nigra* is resistant to canker.

Chrysanthemum eelworms

▶ **SEE** page 82

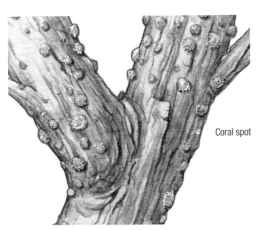
Coral spot

Coral spot

■ **PLANTS AT RISK** Many trees and shrubs, including acers, maples, magnolias and redcurrants.
■ **RECOGNITION** Rashes of pink or coral-red spots on dead twigs. Coral spot is a fungus that can kill trees or shrubs if it is able to enter living shoots. Affected plants may recently have been transplanted or suffered from drought or waterlogging.
■ **DANGER PERIOD** Any time in plants under stress.
■ **TREATMENT** Cut out and burn all dead wood and prune 10–15cm below the diseased area. **Organic advice** Identify and rectify any cultural problems. Avoid injury to bark that would allow the disease to enter.

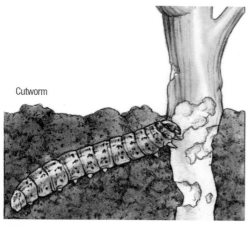

Cutworm

Frost damage

Cutworms

- **PLANTS AT RISK** Lettuces, other vegetables and some young ornamental annuals.
- **RECOGNITION** Stems often eaten at soil level by fat green or grey-brown caterpillars which curl up into a C shape when picked up.
- **DANGER PERIOD** Dry spells from early spring to late summer.
- **TREATMENT** Eliminate weeds, which attract cutworms. **Organic advice** Fork over the soil in winter to expose the pests to predators, particularly blackbirds. Protect young plants with collars made from the cardboard insides of toilet rolls, cans or lengths of drainpipe. Push the collar into the soil around the plant. At the first sign of damage, search in the soil and destroy any caterpillars.

Fire blight

▶ **SEE** page 55

Frost damage

- **PLANTS AT RISK** Young leaves and shoots on most species of flowering plants, shrubs and trees.
- **RECOGNITION** Cracking of tree trunks, blackening and shrivelling of leaves, and blackened flower centres and buds.

- **DANGER PERIOD** Mid winter and spring.
- **TREATMENT** Cut out frostbitten shoots to prevent entry of fungi. Protect plants from further damage by covering them with paper, straw or sacking. Harden off tender plants before planting out.

Gumming

- **PLANTS AT RISK** Cherries and other prunus species.
- **RECOGNITION** Gum, exuding on the surface of branches and trunks, gradually hardens – the result of unsuitable soil conditions or malnutrition.
- **DANGER PERIOD** Any time, but at its worst in summer.
- **TREATMENT** Gumming should stop with good feeding, mulching and watering. Gum may have to be removed in order to cut out dead wood beneath. Do this in October.

Gumming

Honey fungus
(Bootlace fungus)

■ **PLANTS AT RISK** Most trees and shrubs are vulnerable to this fungus. Common among rotting tree stumps, some herbaceous perennials and some bulbs.

■ **RECOGNITION** Toadstools at soil level at the base of the trunk. White fan-shaped growths of fungus occur beneath the bark of roots and at soil level. Black 'bootlace' threads on diseased roots spread infection.

■ **DANGER PERIOD** Autumn.

■ **TREATMENT** Destroy dead or dying plants and as many of their roots as possible. The 'bootlaces' do not always spell disaster for your garden: many species of the fungus are not invasive.

Honey fungus

Lichens
▶ **SEE** page 122

Mustard beetles
▶ **SEE** page 41

Rabbits

■ **PLANTS AT RISK** Many plants, particularly lettuces.

■ **RECOGNITION** Chewed or damaged plants. On young trees, bark can be stripped just above soil level.

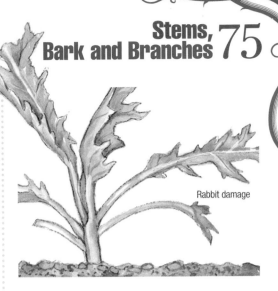

Rabbit damage

■ **DANGER PERIOD** Growing season.

■ **TREATMENT** Protect plants with high wire-mesh fencing, buried 30cm below soil level to prevent burrowing.

Smuts
▶ **SEE** page 48

Spur blight

■ **PLANTS AT RISK** Raspberries and loganberries.

■ **RECOGNITION** Canes develop purple to silver blotches, spotted with black, caused by a fungal disease. The spores spread the disease to healthy plants.

■ **DANGER PERIOD** Spring and summer.

■ **TREATMENT** After fruiting, cut out and destroy infected canes. When new canes are a few centimetres high, spray with penconazole or mycobutanil, and repeat 3 or 4 times at fortnightly intervals.

Spur blight

Squirrels

■ **PLANTS AT RISK** Many shrubs and trees, including ripe fruits. Also bulbs and corms.

■ **RECOGNITION** Bark stripped from trees, fruits damaged and holes in the garden where bulbs and corms have been dug up.

■ **DANGER PERIOD** Between autumn and spring, particularly where gardens adjoin woodland.

■ **TREATMENT** Dust bulbs and corms with proprietary rodent repellent before planting. Protect valuable crops with a covering of netting.

Squirrel damage

Stem canker

■ **PLANTS AT RISK** Conifers.

■ **RECOGNITION** The red fruiting bodies of a fungal disease appear on dead bark. The fungus enters through wounds caused by frost, damage or injury to cause die-back of shoots.

■ **DANGER PERIOD** After frost.

■ **TREATMENT** Cut out all infected shoots right back to healthy wood.

Stem canker

Stem rot

■ **PLANTS AT RISK** Tomatoes, godetias, lobelias and carnations, each of which is affected by a different disease.

■ **RECOGNITION** Rotting of stems – but no obvious fungal growth – is visible on the affected areas.

■ **DANGER PERIOD** Growing season.

■ **TREATMENT** Where possible, cut out and burn all diseased tissue, and spray plants with mancozeb. Alternatively, destroy diseased plants.

Stem rot

Witches' brooms

■ **PLANTS AT RISK** Prunus species and birches.
■ **RECOGNITION** Clusters of erect shoots, growing abnormally from a single point, on infected branches. The condition is caused by a fungal disease.
■ **DANGER PERIOD** At any time throughout the life of the tree.
■ **TREATMENT** Cut off the affected branch to a point 15cm below the broom.

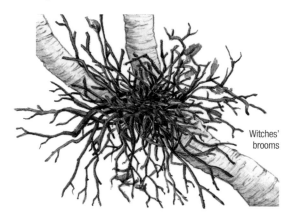

Witches' brooms

Woolly aphids

▶ **SEE** page 70

Azalea gall

Azalea gall

■ **PLANTS AT RISK** Small-leaved rhododendrons, including pot-grown *Rhododendron simsii* and Kurume azaleas.

■ **RECOGNITION** Leaves and flowers are replaced by red or pale green fleshy galls, which later produce a white floury coating of fungal spores. The disease is spread by air or by insects which carry the spores on to the plants.

■ **DANGER PERIOD** Growing season.

■ **TREATMENT** Remove and burn the galls before they turn white and produce fresh spores. In the greenhouse control spore-bearing insects, such as aphids. Spray severe attacks before the leaves curl, using Bordeaux mixture or other copper fungicide.

Crown gall

■ **PLANTS AT RISK** Many species, including soft and tree fruit, vegetables, shrubs and herbaceous perennials.

■ **RECOGNITION** Hard or soft galls, sometimes in a chain along a root or shoot and often occurring in wet soil, are the result of bacteria entering through wounds.

■ **DANGER PERIOD** Growing season.

■ **TREATMENT** Destroy infected plants and provide suitable drainage to prevent soil from becoming waterlogged. Avoid damage when planting and during cultivation.

Crown gall

Gall midges

■ **PLANTS AT RISK** Many cultivated species.

■ **RECOGNITION** Disfiguring galls appear on stems, leaves and flowers. The galls are caused by tiny red, white or yellow maggots.

■ **DANGER PERIOD** Late winter, early spring.

■ **TREATMENT** Burn infected tissues in early spring.

Gall midges

Gall mites

Gall mites

■ **PLANTS AT RISK** Many deciduous trees, including elm, lime, maple and sycamore. Also blackcurrants.

■ **RECOGNITION** Small galls on the upper surface of the leaves caused by mites. The galls are harmless to trees. The mites cause enlarged buds and transmit reversion virus in blackcurrants.

■ **DANGER PERIOD** Spring. From February and March on blackcurrants.

■ **TREATMENT** None for trees but remove and burn affected buds on blackcurrants in early March.

Gall wasps

■ **PLANTS AT RISK** Oaks, and some species of roses and willows.

■ **RECOGNITION** Solitary or numerous galls grow out of leaves. The galls resemble peas, cherries or silk buttons. Caused by

Gall wasps

tiny wasps living in plant tissues.

■ **DANGER PERIOD** May to October.

■ **TREATMENT** Remove and destroy the galls if possible; however, damage to the plant is rarely serious, so gall wasps are not a real cause for concern.

Leafy gall

■ **PLANTS AT RISK** Many species, but in particular sweet peas, dahlias, carnations, gladioli, pelargoniums, chrysanthemums and strawberries.

■ **RECOGNITION** Abortive shoots, often flattened with thickened and distorted leaves, develop on the plant at soil level due to bacterial infection.

■ **DANGER PERIOD** During propagation and the growing season.

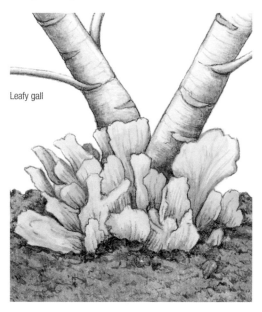

Leafy gall

■ **TREATMENT** Remove and burn all infected plants. Wash your hands after handling infected material and wash tools in tri-sodium orthophosphate. When replanting, choose a non-susceptible plant. Always grow susceptible plants in clean containers and sterilised compost.

Birds
▶ **SEE** page 98

Blindness

■ **PLANTS AT RISK** Narcissi and tulips, especially those grown in containers.
■ **RECOGNITION** Shoots fail to develop flower buds – or those which do turn brown and wither – due to waterlogged or excessively dry roots.
■ **DANGER PERIOD** Hot summers.
■ **TREATMENT** The cause may be water-logged or very dry roots, or BASAL ROT (p. 112). Blindness can also be caused by shallow planting. If so, allow container plants to die down, then store the bulbs in a well-lit cool shed over summer and replant at the correct depth in autumn.

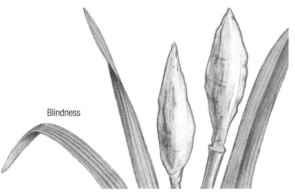

Blindness

Bud-drop

■ **PLANTS AT RISK** Camellias, sweet peas, wisterias and many houseplants.
■ **RECOGNITION** Buds drop before flowering.
■ **DANGER PERIOD** Growing season.
■ **TREATMENT** Nothing can save falling buds. Prevent bud-drop by mulching with organic matter. Water well in dry periods – for camellias particularly June-September. Bud-drop is caused by a shortage of water, sometimes in autumn when you may not think plants need watering.

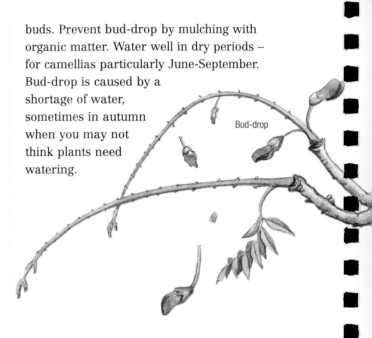

Bud-drop

Capsid bugs
▶ **SEE** page 39

Chrysanthemum eelworms
▶ **SEE** page 82

Frost damage
▶ **SEE** page 74

Leaf rot
▶ **SEE** page 58

Reversion

■ **PLANTS AT RISK** Blackcurrants.
■ **RECOGNITION** Flower buds are hairless and a bright magenta instead of the normal dull grey. Despite vigorous shoots,

Reversion

of fungus spores which are transmitted by rhododendron leafhoppers.

■ **DANGER PERIOD** Between October and December.

■ **TREATMENT** Control the rhododendron leafhoppers (see p. 38). Destroy all affected buds and spray with Bordeaux mixture just before flowering and at monthly intervals.

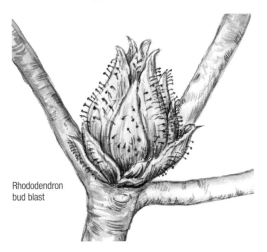

Rhododendron bud blast

the leaves are also smaller and have fewer lobes than normal. The disease is caused by a virus.

■ **DANGER PERIOD** Early summer.

■ **TREATMENT** Control the gall mites that spread the virus (see p. 79). Destroy badly diseased bushes and replace with plants which are certified as free of virus disease.

Rhododendron bud blast

■ **PLANTS AT RISK** Evergreen rhododendron species and hybrids.

■ **RECOGNITION** Buds turn grey or brown. In spring, black bristle-like structures appear on the infected buds, bearing a pinhead

Rhododendron leafhoppers

▶ **SEE** page 38

Tarsonemid mites

▶ **SEE** page 48

Virus diseases VIRUSES ARE MICROSCOPIC PARTICLES, CAPABLE OF CAUSING DISEASE IN LIVING CELLS, AND CAN ENTER PLANT TISSUES ONLY THROUGH WOUNDS.

Viruses may be transmitted through the air, soil, fungi, seeds, vegetative grafting, propagation and by handling. Most viruses, however, are spread by insects, such as aphids, eelworms, leafhoppers, mites, thrips and whiteflies. Symptoms of a virus disease vary. There may be colour variation in the leaves, stems, flowers and tubers, or the plant may display distorted foliage, wilting, stunted growth and tissue decay. A plant may suffer from one or many viruses, with a different combination of viruses producing a number of different symptoms.

There is no chemical treatment for virus-infected plants, and any that are severely diseased should be destroyed. Use only healthy plants for propagation and where possible buy only new stock certified as being free of viruses. Good garden hygiene and the control of the insects that carry the viruses both help to prevent virus diseases from entering the garden.

Blotch

Blotch

■ **PLANTS AT RISK** Delphiniums and hellebores.

■ **RECOGNITION** Flowers develop black blotches.

■ **DANGER PERIOD** Wet seasons.

■ **TREATMENT** Cut out and burn affected parts. Spray plants with mancozeb at first sign of infection.

Chrysanthemum eelworms (Leaf and bud eelworms)

■ **PLANTS AT RISK**
Primarily chrysan-themums, but also peonies, callistephus, calceolarias and other ornamental plants.

Chrysanthemum eelworms

■ **RECOGNITION** Buds damaged, blooms distorted and stems scarred. Caused by microscopic worms. Severe infestation can kill plants.

■ **DANGER PERIOD** July to September.

■ **TREATMENT** There is no safe treatment for chrysanthemum eelworms. Dig up and burn all infected plants.

Chrysanthemum virus diseases

Chrysanthemum virus diseases

■ **PLANTS AT RISK** Most varieties of chrysanthemum, grown outdoors and under glass.

■ **RECOGNITION** Distorted blooms, break-up or greening of the flower colour, leaf mottling or stunted growth – the result of viruses transmitted by pests, including aphids.

■ **DANGER PERIOD** Growing season.

■ **TREATMENT** Destroy all the affected plants. Prevent further infection by washing your hands in hot soapy water and sterilising any equipment that has been used in tri-sodium orthophosphate.

Fire blight
▶ **SEE** page 55

Frost damage
▶ **SEE** page 74

Grey mould
▶ **SEE** page 56

Narcissus fire

Narcissus fire
■ **PLANTS AT RISK** Narcissi – particularly in the West Country.
■ **RECOGNITION** Rotting flowerheads caused by fungal disease, together with GREY MOULD (p. 56) spreading to the leaves.
■ **DANGER PERIOD** Wet seasons.
■ **TREATMENT** Burn infected leaves to stop fungus overwintering. Remove plants showing infection in spring. Spray the remainder when 2.5cm high with mancozeb or penconazole. Repeat at 10-day intervals until the flower-buds emerge from the leaf sheaths.

Petal blight
■ **PLANTS AT RISK** Chrysanthemums and, occasionally, cornflowers, dahlias and heads of globe artichokes.

Petal blight

■ **RECOGNITION** Dark, water-soaked spots spread on florets until flowers rot.
■ **DANGER PERIOD** Wet seasons.
■ **TREATMENT** Destroy affected flowers. Spray with mancozeb. Prevent disease in greenhouse by keeping humidity low.

Ray blight
■ **PLANTS AT RISK** Chrysanthemums in the greenhouse.
■ **RECOGNITION** Dark, water-soaked spots or blotches on petals cause flowers to rot.
■ **DANGER PERIOD** Spreads rapidly in humid conditions when the temperature reaches 16°C.
■ **TREATMENT** Reduce humidity and remove all infected plants. Change the soil.

Ray blight

Thrips

Thrips

■ **PLANTS AT RISK** Roses, carnations, privets, gladioli and some other species. Also peas and onions.

■ **RECOGNITION** Fine silvery flecks on petals and foliage, caused by tiny thrips. Severe infestations can discolour and kill the flowers.

■ **DANGER PERIOD** Early summer to early autumn, particularly in hot, dry weather.

■ **TREATMENT** Treat corms before storage – and again before planting – with derris dust. If symptoms appear, spray the plants with fatty acid sprays or imidacloprid.

Organic advice Keep the plants growing strongly by giving them good conditions. Do not let the soil dry out.

Tulip fire

■ **PLANTS AT RISK** Tulips.

■ **RECOGNITION** Small brown spots on the petals, which later become covered in GREY MOULD (p. 56). Caused by a fungal disease. Bulbs rot and bear small black fungal growths.

■ **DANGER PERIOD** Flowering time, particularly in cold wet weather; bulbs shortly before or after planting.

■ **TREATMENT** Destroy all diseased plants and rotting or fungus-bearing bulbs. If possible, choose a fresh site each year, especially after the disease has appeared. Spray with penconazole or mancozeb when leaves are about 5cm high and repeat at 10-day intervals until flowering.

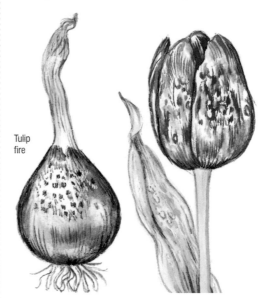

Tulip fire

Weevils

▶ **SEE** page 38

Aster wilt

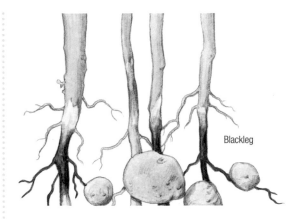

Blackleg

Aster wilt

■ **PLANTS AT RISK** Primarily asters of the *novi-belgii* species.

■ **RECOGNITION** The leaves turn yellow and wilt, then die. The disease is caused by a fungus that inhabits the rootstock, transmitting a poison through the sap.

■ **DANGER PERIOD** Growing season.

■ **TREATMENT** Destroy infected plants and – since the disease is soil-borne – propagate from healthy stock on a fresh site.

Blackleg

■ **PLANTS AT RISK** Pelargoniums and potatoes.

■ **RECOGNITION** Affected tissues become soft, the leaves turn yellow and the cutting or plant dies – the result of a black rot developing at the base of the stem. The rot is caused by fungi or bacteria.

■ **DANGER PERIOD** Soon after pelargonium cuttings are taken. In June for potatoes. A problem on cold wet soil.

■ **TREATMENT** Destroy severely affected plants. No treatment for potatoes other than to plant healthy seed tubers and to choose less susceptible varieties. Protect pelargoniums with sterilised compost. Maintain strict greenhouse hygiene and water carefully with mains water.

Brown core

■ **PLANTS AT RISK** Primulas.

■ **RECOGNITION** Plants wilt and can easily be lifted due to the roots rotting back from the tips.

■ **DANGER PERIOD** Growing season.

■ **TREATMENT** The disease only affects plants that are set too deeply in the soil or are overcrowded. Burn affected plants and avoid growing primulas on the infected site for several years.

Brown core

Cabbage root flies

■ **PLANTS AT RISK** Recently transplanted brassicas, especially cabbages, cauliflowers and brussels sprouts. Radishes, turnips, swedes and wallflowers.

■ **RECOGNITION** Outer leaves wilt and develop a purple/blue/red tinge. Young plants wilt and are easily pulled out of the ground. Established plants may survive but give a smaller yield. Small white maggots feed on the roots.

■ **DANGER PERIOD** April to August.

■ **TREATMENT** **Organic advice** Protect transplanted seedlings with a 13cm square of rubbery carpet underlay, slit to the centre. Place it on the soil around the stem. Cover crop with horticultural fleece. Alternatively, plant brassicas next to strong-smelling crops, such as onions or garlic.

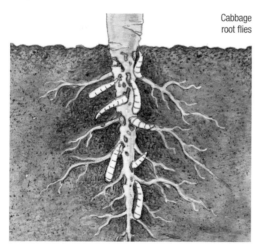

Cabbage root flies

Callistephus wilt

▶ **SEE ALSO** FUSARIUM WILT (p. 90)

■ **PLANTS AT RISK** Callistephus.

■ **RECOGNITION** Blackened stems with white or pink fungus outgrowths. Prevalent on wet, badly drained soils.

■ **DANGER PERIOD** Growing season.

■ **TREATMENT** Remove and burn infected

Callistephus wilt

plants. Improve drainage by adding coarse grit to soil or installing a drainage system.

Cane blight

■ **PLANTS AT RISK** Raspberries.

■ **RECOGNITION** Leaves on fruiting canes wilt and wither in summer, and canes may snap in the wind. The canes have dark patches at ground level and the bark splits or cracks.

■ **DANGER PERIOD** Growing season.

■ **TREATMENT** Cut out all diseased canes to below soil level and burn. Disinfect knife immediately. Spray new canes with Bordeaux mixture. Handle canes with care to prevent damage. Do not transplant infected canes to a new site. **Organic advice** Tie in canes to prevent wind-rock, which can cause damage that becomes an entry point for the disease.

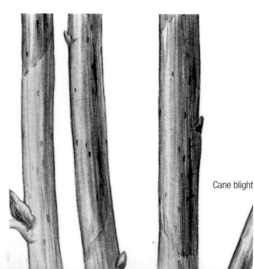

Cane blight

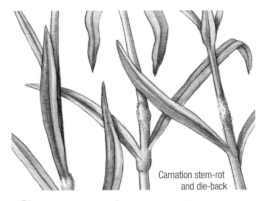

Carnation stem-rot
and die-back

Carnation stem-rot and die-back

■ **PLANTS AT RISK** All species.

■ **RECOGNITION** Rotting caused by soil-borne fungus that enters the stems via wounds.

■ **DANGER PERIOD** During wet periods and on poorly drained soil.

■ **TREATMENT** Spray stock plants with carbendazim. Use only sterilised composts for potting.

Chafer beetles

■ **PLANTS AT RISK** Many annuals and perennials, tubers and stems of fruit and vegetables.

■ **RECOGNITION** Top growth wilts and dies as the fat, white chafer beetle larvae eat the roots. The distinctive C-shaped larvae are 4cm long with three pairs of legs and a brown head.

■ **DANGER PERIOD** May and June.

■ **TREATMENT** **Organic advice** Mainly a problem of newly cultivated land. Numbers are quickly reduced by cultivation and weed control.

Clematis wilt

■ **PLANTS AT RISK** Clematis, especially large-flowered varieties.

■ **RECOGNITION** Upper parts of shoot wilt, with young leaves wilting first and leaf stalks blackening where they join the blade. Caused by fungus.

■ **DANGER PERIOD** Early spring and through the growing season.

■ **TREATMENT** Cut wilted shoots back to below soil level. Drench shoots with fungicide when the first damage is seen and again in spring. **Organic advice** Plant new clematis plants 15cm deeper than their level in the pot. Avoid wounding stems when planting, and tie stems to their support in the growing season to prevent damage from wind or accidental contact. Clear up all diseased material and burn it.

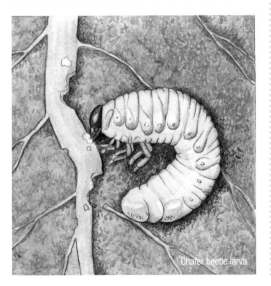

Chafer beetle larva

Clematis
wilt

Club root

- **PLANTS AT RISK** Brassicas, wallflowers, stocks, turnips and radishes.
- **RECOGNITION** Leaves of brassicas wilt and turn red, purple or yellow; roots become swollen and distorted. Plants wilt on hot days, but may recover if watered.
- **DANGER PERIOD** Throughout the growing season.
- **TREATMENT** Once soil is infected, little can be done. Prevent the disease by liming the soil in winter to keep pH at 7–7.5. On badly drained soil make raised beds. Never bring in plants that may have the disease; raise your own from seed in sterilised compost. Kale, winter broccoli and spring cabbage grow better in infected soil than other brassicas. There are a few resistant types.

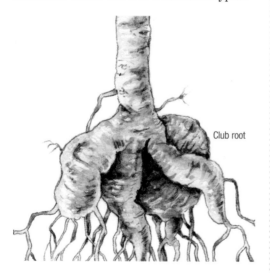

Club root

Collar rot

- **PLANTS AT RISK** Plants grown in greenhouses.
- **RECOGNITION** Plants collapse at or just above soil level due to organisms rotting the tissue.
- **DANGER PERIOD** Any time.
- **TREATMENT** Remove all dead and decaying tissues, and dust with Bordeaux powder. Repot plants in a lighter, sterilised compost and water carefully.

Collar rot

Crown rot of rhubarb

- **PLANTS AT RISK** Rhubarb.
- **RECOGNITION** Leaves become spindly and discoloured and die early. The main bud rots, followed by the whole of the crown, due to bacterial disease. Most common in wet soil.
- **DANGER PERIOD** Growing season.
- **TREATMENT** Dig up and burn entire plant and never plant rhubarb on the same site again.

Crown rot of rhubarb

Damping-off

- **PLANTS AT RISK** Seedlings sown under glass in crowded and wet conditions.
- **RECOGNITION** Seedlings rot at ground level, topple over and die – killed by a parasitic fungus.

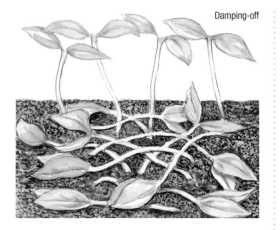

Damping-off

■ **DANGER PERIOD** From seed-sowing to emergence of third pair of true leaves.
■ **TREATMENT** Remove and destroy dead seedlings. Water seed boxes with copper or Cheshunt compound. To prevent future trouble, sow thinly, use a sterilised seed compost, water carefully and ensure good ventilation. Always use clean containers, and never use rainwater from a butt unless the butt is cleaned regularly.

Dianthus leaf spot
▶ **SEE** page 54

Die-back
■ **PLANTS AT RISK** Mainly fruit trees and shrubs.
■ **RECOGNITION** Shoots die back from the tips, often killing larger branches. Foliage turns brown or yellow and withers.

Die-back

Damage may be due to frost, fungal infection or PHYSIOLOGICAL DISORDERS (SEE p. 61).
■ **DANGER PERIOD** Any time.
■ **TREATMENT** Cut out all infected areas back to healthy wood and burn the cuttings. Treat identifiable disease appropriately; if unidentifiable, improve cultural conditions.

Fire blight
▶ **SEE** page 55

Drought-resistant plants

PLANT A SELECTION OF DROUGHT-TOLERANT PLANTS and, even in dry summers, you can reduce watering to a minimum. Most of these are undemanding and will do well in dry soil.

TREES AND SHRUBS
Berberis, Buddleja, Ceanothus, Choisya, Cistus, Cotoneaster, Deutzia, Escallonia, Euonymus, Fatsia, Genista, Hebe, Ilex, Juniperus, Lavandula, Ligustrum, Phlomis, Potentilla, Ribes, Robinia, Rosmarinus, Rubus, Santolina, Senecio, Sorbaria, Symphoricarpos, Syringa, Tamarix

PERENNIALS
Acanthus, Achillea, Anchusa, Artemesia, Campanula, Dianthus, Echinops, Eryngium, Euphorbia, Gaillardia, Gypsophila, Linum, Malva, Nepeta, Platycodon, Rudbeckia, Stachys, Stipa

BULBS, CORMS AND TUBERS
Allium, Alstroemeria, Chionodoxa, Crocosmia, Cyclamen, Gladiolus, Hyacinthus, Nerine, Ornithogalum, Tulipa

ANNUALS
Antirrhinum, Calendula, Campanula, Clarkia, Cosmos, Dianthus, Helianthus, Nigella, Papaver, Pelargonium, Petunia, Scabiosa, Tagetes

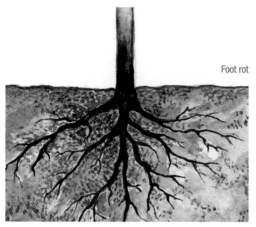

Foot rot

Foot rot

■ **PLANTS AT RISK** Tomatoes, bedding plants, sweet peas, peas and beans.
■ **RECOGNITION** Stem bases turn black and rot, and the roots usually die due to various fungi.
■ **DANGER PERIOD** Growing season.
■ **TREATMENT** Rotate vegetable and bedding plants, and always use sterile compost for pot plants. Water bedding plants in seed boxes with Cheshunt compound when they are put in their planting holes. If necessary, repeat at weekly intervals.

Fusarium wilt

▶ **SEE ALSO CALLISTEPHUS WILT (p. 86)**
■ **PLANTS AT RISK** Carnations, dianthus species, dwarf and runner beans, garden peas and sweet peas.

Fusarium wilt

■ **RECOGNITION** Leaves and sometimes stems become discoloured and plants wilt due to several species of fungus.
■ **DANGER PERIOD** Growing season.
■ **TREATMENT** Remove and burn the infected plants. Sterilise the soil or grow susceptible plants on a fresh site each year. Propagate carnations and dianthus species from healthy plants only. Drench with carbendazim, according to the instructions on the label.

Gladiolus yellows

▶ **SEE** page 56

Heather die-back

Heather die-back

■ **PLANTS AT RISK** Calluna and erica.
■ **RECOGNITION** The foliage develops a grey tinge, then wilts, turns brown and dies.
■ **DANGER PERIOD** Throughout the growing season in badly drained soil.
■ **TREATMENT** No cure. Dig up and burn infected plants. Keep remaining plants healthy by applying an acid-formulation fertiliser in April, and mulching with leaf-mould, shredded bark or woodchips. Improve drainage.

Ink
disease

Ink disease

■ **PLANTS AT RISK** Bulbous irises, lachenalias, tritonias and crocosmias.
■ **RECOGNITION** As the young leaves show through the ground, they have yellow streaks which become black blotches. If the leaves develop, they may turn red and wither. Diseased bulbs have black crusty patches on their outer scales.
■ **DANGER PERIOD** Wet season.
■ **TREATMENT** Remove and burn all infected bulbs.

Leatherjackets
(Crane-fly larvae)

■ **PLANTS AT RISK** Brassicas as well as some vegetables and ornamental plants. Young plants and lawns are particularly susceptible.
■ **RECOGNITION** Plants turn yellow, wilt and sometimes die as the roots are attacked by

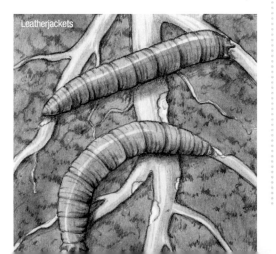

Leatherjackets

fat, grey-brown, legless grubs. Large numbers of grubs on a lawn will cause yellow patches in dry spells.
■ **DANGER PERIOD** April to June.
■ **TREATMENT** On lawns, water soil with imidacloprid **Organic advice** Fork over infested soil thoroughly and frequently to expose pests to birds. On lawns, water the grass well then cover it with sacking or thick cardboard overnight. The leatherjackets will come to the surface and can be swept up or left for the birds. On cultivated land, put grass mowings under the cover, leave for two days, then uncover.

Lily disease
▶ **SEE** page 60

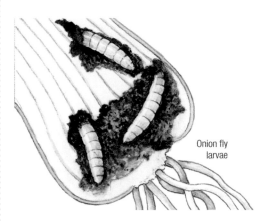

Onion fly
larvae

Onion flies

■ **PLANTS AT RISK** Young plants of onions, shallots and leeks.
■ **RECOGNITION** Adult flies lay eggs on the leaves. The larvae hatch out and tunnel into the plants. Leaves turn yellow and wither, and the stems and bulbs rot. Small white maggots found in roots and bulbs.
■ **DANGER PERIOD** June and July.
■ **TREATMENT** Burn infested plants and dig over the ground to expose insects' dormant pupae to the winter cold and to birds, which will eat them.

Pansy sickness

■ **PLANTS AT RISK** Violas.
■ **RECOGNITION** Roots and stems are attacked by soil-borne organisms causing plants to collapse.

Pansy sickness

■ **DANGER PERIOD** Growing season.
■ **TREATMENT** Remove and destroy diseased plants and their roots. Water seedlings and established plants with Cheshunt compound, repeating at weekly intervals if necessary. Choose a new site every year.

Peony wilt
(Peony blight)

■ **PLANTS AT RISK** Peonies.
■ **RECOGNITION** Grey mould forms on the stems at soil level, which causes them to

Peony wilt

collapse. The fungus then spreads to form brown blotches on the leaves of neighbouring plants.
■ **DANGER PERIOD** Growing season.
■ **TREATMENT** Cut out all affected shoots to below soil level, and burn them. Dust the crowns of the plants with dry Bordeaux powder. Spray with mancozeb soon after the leaves appear.

Petunia wilt

■ **PLANTS AT RISK** Petunias, zinnias, salpiglossis and other bedding plants.
■ **RECOGNITION** Plants wilt, often as they are about to flower, and the base of the stems may become discoloured.

Petunia wilt

■ **DANGER PERIOD** Growing season.
■ **TREATMENT** Burn all diseased plants. Rotate bedding plants so that vulnerable types are grown on a fresh site each year.

Phytophthora root rot

■ **PLANTS AT RISK** Trees and shrubs. Most common on Lawson cypress, azaleas, rhododendrons, beech, heathers, limes, apples, prunus and yews.
■ **RECOGNITION** Small, sparse yellow foliage, with partial die-back often up one side of the plant. The whole plant may die.

Phytophthora
root rot

Sometimes only recognised when new
growth fails to appear in spring.
■ **DANGER PERIOD** Warm wet periods.
■ **TREATMENT** None. Only obtain plants
from a reputable nursery. Avoid poorly
drained soil. Never transplant from an
infected area onto clean soil.

Potassium deficiency
▶ **SEE** page 62

Powdery mildew
▶ **SEE** page 62

Rhizome rot
■ **PLANTS AT RISK** Rhizomatous irises.
■ **RECOGNITION** Leaf fan collapses at soil

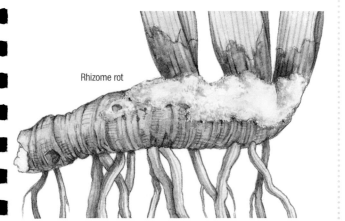

Rhizome rot

level due to the soft, yellow and
foul-smelling rot at its growing point.
■ **DANGER PERIOD** Any time, but mostly in
wet weather.
■ **TREATMENT** Destroy badly infected plants.
On less severe outbreaks, cut out the
rotting parts and dust the wounds with
Bordeaux mixture.

Root aphids
▶ **SEE** page 111

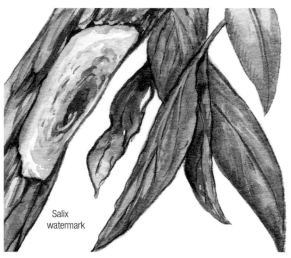

Salix
watermark

Salix
watermark
■ **PLANTS AT RISK** Willows – most common
on *Salix alba* varieties.
■ **RECOGNITION** Leaves turn red and
wilt, then die but remain on the tree.
Infection can be confirmed by cutting
through an affected branch. A watery
reddish brown or brown-black stain
will be found on the wood.
■ **DANGER PERIOD** Growing season.
■ **TREATMENT** Remove and burn all infected
plants. Sterilise tools that have been used
with tri-sodium orthophosphate.

Sclerotinia disease

Seedling blight

Sclerotinia disease

■ **PLANTS AT RISK** Herbaceous perennials, bulbs, corms, tubers – including stored root crops, especially carrots and parsnips.

■ **RECOGNITION** Plants may wilt suddenly, develop yellow basal leaves and collapse where the disease entered. Diseased tissue has a white fluffy mass containing large black fungal growth. The roots eventually soften and decay.

■ **DANGER PERIOD** During growing season and when crops are in store.

■ **TREATMENT** Burn all diseased plants to prevent risk of the infection spreading. **Organic advice** Use a minimum four-year rotation. Control weeds in the garden, on which the disease may overwinter.

Seedling blight

■ **PLANTS AT RISK** Zinnias.

■ **RECOGNITION** Leaves develop red-brown spots with grey centres; the stems develop brown canker-like areas; and the seedlings collapse and die – all due to a fungal disease.

■ **DANGER PERIOD** While the plant is small.

■ **TREATMENT** Once the symptoms have appeared, the affected plants must be burnt. Protect seedlings in boxes with a copper-containing fungicidal spray.

Silver leaf
▶ **SEE** page 64

Stem rot
▶ **SEE** page 76

Verticillium wilt

■ **PLANTS AT RISK** Cotinus, rhus and acer, tomatoes and carnations in greenhouses, Michaelmas daisies.

■ **RECOGNITION** The leaves of one or two shoots wilt, and affected branches eventually die back. The disease is caused by a soil-borne fungus entering wounds. On tomatoes all leaves wilt, but may recover overnight.

Verticillium wilt

■ **DANGER PERIOD** Growing season.

■ **TREATMENT** On trees and shrubs cut back affected branches to living tissue. If trouble recurs, lift and burn the plant. Destroy diseased greenhouse plants, and sterilise the greenhouse at the end of the season. Take tip cuttings of diseased Michaelmas daisies, and burn old plants. Carbendazim gives some control.

Vine weevils

■ **PLANTS AT RISK** Many, especially those grown in containers. Cyclamen, primulas and begonias are particularly susceptible.

■ **RECOGNITION** Plants wilt and collapse suddenly. Fat, white, brown-headed larvae are found in the soil around the roots. Adult weevils eat notches out of the leaves. Most damage is done under glass.

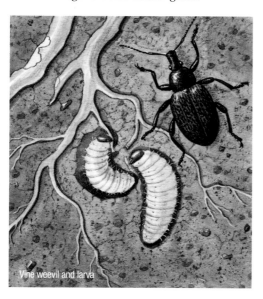
Vine weevil and larva

■ **DANGER PERIOD** Most of the year.

■ **TREATMENT** Water soil thoroughly with thiacloprid. **Organic advice** Check the compost of newly bought plants. Protect pots and greenhouse staging with a band of non-drying glue, which can be bought at some garden centres. Adult vine weevils cannot fly; they must walk up the side of pots. Use a parasitic nematode *Heterorhabditis*.

Violet root rot

■ **PLANTS AT RISK** Primarily carrots, parsnips and asparagus, but also some ornamentals and fruit crops.

■ **RECOGNITION** Leaves turn yellow and die as purple-violet threads of fungus attack the roots.

■ **DANGER PERIOD** Growing season.

■ **TREATMENT** Lift and burn diseased plants. Isolate the infected area by sinking pieces of rigid polythene or corrugated iron 30cm deep for the rest of the life of the bed. Avoid growing vulnerable plants there for several years. **Organic advice** Avoid planting crops in wet, acid soils.

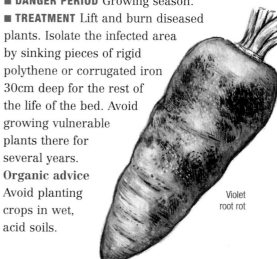
Violet root rot

Winter killing

■ **PLANTS AT RISK** Mainly wallflowers.

■ **RECOGNITION** Sideshoots either die back or the plant dies. Caused by FROST DAMAGE (p. 74) and GREY MOULD (p. 56).

■ **DANGER PERIOD** Late winter to late spring.

■ **TREATMENT** Set out plants early so that they are fully established before winter and carry out good cultivation.

Winter killing

Wire stem

■ **PLANTS AT RISK** Brassicas and other seedlings.

■ **RECOGNITION** The base of the stem turns brown and shrinks, and easily breaks. Seedlings die or become stunted, both due to a fungal disease.

■ **DANGER PERIOD** During early growth.

■ **TREATMENT** Destroy diseased plants. Raise seedlings in sterilised compost.

Wire stem

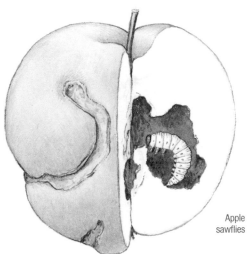

Apple
sawflies

Apple sawflies

- **PLANTS AT RISK** Apples and plums.
- **RECOGNITION** Small ant-like flies lay their eggs in the blossom. The newly hatched caterpillars burrow through to the cores of the young fruit which fall prematurely or develop scars as they grow.
- **DANGER PERIOD** May and June.
- **TREATMENT** Spray thoroughly at weekly intervals when the petals start to fall, using derris or pyrethins.

Organic advice Spray with derris liquid – usually at petal-fall.

Apple scab

- **PLANTS AT RISK** Only trees of the genus *Malus*.
- **RECOGNITION** Small blister-like pimples on young shoots and fallen leaves. The blisters later burst the bark and form ring-like cracks or scabs.
- **DANGER PERIOD** Growing season, but particularly after a wet May.
- **TREATMENT** Cut out and burn diseased shoots in winter. Spray regularly with myclobutanil or mancozeb (except on sulphur-shy varieties). **Organic advice** Pick up fallen apple leaves in autumn, or run a mower over them. Worms will take the pieces into the soil, burying the spores. Grow resistant varieties.

Apple scab

Preventing fruit tree problems

Trap winter moths by fixing grease bands (see right) round the trunks of apple, pear, plum and cherry trees. A barrier of grease applied in October will trap wingless female moths as they climb up the tree in autumn and winter to lay their eggs.

- Spread the grease onto the bark or onto a strip of paper 10cm wide, fixed in a continuous band around the trunk; prepared grease bands are also available.
- Tree stakes provide an alternative route for the moths, so grease these too.
- Check occasionally that the bands are still sticky, and leave in place until April.

Birds

■ **PLANTS AT RISK** Plums, gooseberries, pears, flowering cherries and forsythia damaged by bullfinches; ripening apples and pears by blackbirds and tits; many plants attacked by sparrows; brassica seedlings and fruit by wood pigeons.
■ **RECOGNITION** Flower buds, ripening fruit and seedlings eaten.
■ **DANGER PERIOD** November to May.
■ **TREATMENT** Cover plants with netting. Use bird scarers such as black cotton or string stretched between canes above plants, strips of foil or old compact discs hung from lines or canes, and small wind-mills, preferably with a rattle. Change the scarer at least once a week.

Bird damage

Bitter pit

■ **PLANTS AT RISK** Apples.
■ **RECOGNITION** Sunken brown spots develop beneath the skin of the fruit and throughout the flesh.
■ **DANGER PERIOD** Particularly in hot dry summers with acute water shortages. Not apparent until harvesting or in storage.
■ **TREATMENT** Feed, mulch and prevent the soil from drying out. In mid June spray with calcium nitrate – 8 rounded table-spoons to 23 litres of water – and repeat three times at three-weekly intervals.
Organic advice Avoid high applications of nitrogen-rich manures.

Bitter pit

Brown rot

■ **PLANTS AT RISK** All tree fruit.
■ **RECOGNITION** Concentrically ringed browny white fungus appears on fruits originally injured by insects and birds.
■ **DANGER PERIOD** Summer, and in store.
■ **TREATMENT** Destroy infected fruit and cut out dead shoots. Spray fruit with mancozeb before picking to protect in storage. Do not store diseased or damaged fruit. **Organic advice** At end of season remove remaining fruit from the tree. Remove and destroy all plant debris.

Brown rot

Bryobia mites
▶ **SEE** page 52

Capsid bugs
▶ **SEE** page 39

Codling moths

■ **PLANTS AT RISK** Mainly apples but also pears.

■ **RECOGNITION** Entry tunnels through the eye end of fruit where caterpillars eat their way to the core. Damage is rarely noticed until the fruit is cut open.

■ **DANGER PERIOD** June to August.

■ **TREATMENT** Spray in May to June with permethrin or derris, and repeat 3 weeks later to kill young caterpillars before they enter the fruit. **Organic advice** Encourage bluetits into the garden, especially in winter; they are very effective predators of codling moths. Hang pheromone traps in the trees from late spring to late summer to give partial control and to indicate when spraying is needed.

Codling moth larva

Cracking

► **SEE** page 114

Fruit-drop

■ **PLANTS AT RISK** All tree fruit.

■ **RECOGNITION** Fruits drop prematurely while still immature.

■ **DANGER PERIOD** Flowering and just after.

■ **TREATMENT** Ensure that pollinators are in the garden, and feed, mulch and water the

Fruit drop

tree. In cold seasons, fruit-drop is due to poor pollination and nothing can be done.

Gooseberry mildew (American)

■ **PLANTS AT RISK** Gooseberries.

■ **RECOGNITION** White mealy powder, caused by a fungal disease, covers leaves, shoots and fruits which later turn brown. Shoots may become distorted. The fruits are small and tasteless.

■ **DANGER PERIOD** April onwards.

■ **TREATMENT** Remove diseased shoots and burn them immediately. Spray plant with myclobutanil. **Organic advice** A problem when soil is dry and the weather is humid. Mulch plants and keep them well pruned to allow air to circulate. Grow resistant varieties. Do not overfeed with nitrogen-rich fertilisers.

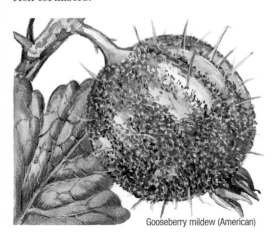

Gooseberry mildew (American)

Gooseberry
mildew (European)

Gooseberry mildew (European)

■ **PLANTS AT RISK** Gooseberries, black-currants and occasionally redcurrants.

■ **RECOGNITION** A light, powdery white covering develops on the upper leaf surface and sometimes on the underside and berries. Widespread on old bushes, but not as damaging as American gooseberry mildew.

■ **DANGER PERIOD** On gooseberries late April; late May to September on currants.

■ **TREATMENT** As for GOOSEBERRY MILDEW (AMERICAN) (p. 99).

Grey mould

▶ **SEE** page 56

Gummosis of cucumber

■ **PLANTS AT RISK** Cucumbers, melons, vegetable marrows and courgettes that are grown in a greenhouse or coldframe.

■ **RECOGNITION** Small grey sunken spots exude sticky liquid. This develops into a velvety dark green mould, and the fruits crack. The disease sometimes appears on the leaves and stems as brownish patches.

■ **DANGER PERIOD** Inadequate ventilation and heat under cool damp conditions.

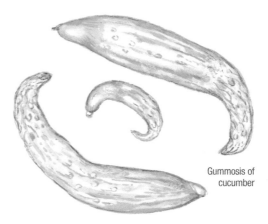

Gummosis of
cucumber

■ **TREATMENT** Burn all diseased fruits. Disinfect greenhouse or coldframe with phenolic compound or a garden disinfectant before planting another crop.

Pear midges

■ **PLANTS AT RISK** Pears.

■ **RECOGNITION** Young fruits swell rapidly, then fail to develop. Black cavities contain small yellowish white larvae. Affected fruit falls in May/June.

■ **DANGER PERIOD** April to May.

■ **TREATMENT** Spray with derris just before the blossoms open. **Organic advice** Pick off affected fruitlets when first seen and burn them. Fork over the soil under the trees in summer and remove old leaves or fruit in autumn.

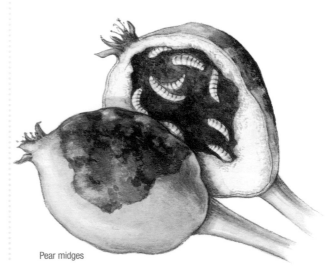

Pear midges

TAKING ACTION TO PREVENT PESTS AND DISEASES IS MORE TIME-EFFECTIVE THAN TRYING TO ERADICATE THEM ONCE THEY HAVE TAKEN HOLD. AN ANNUAL WINTER PRUNE AND CLEAR OUT OF DISEASED PLANT MATERIAL WILL DO MUCH TO IMPROVE THE HEALTH OF FRUIT TREES.

MAINTAINING GOOD CULTURAL HABITS HELPS KEEP PLANTS RESISTANT TO DISEASE and reduces the amount of chemicals needed to control pests. Cultural control involves pruning plants and keeping the garden free of dead matter and debris that can harbour pests and diseases.

WINTER PRUNING GUIDE
A simple winter pruning routine applies to all tree and bush fruits except plums, cherries, peaches and other stone fruits. These are pruned in spring.
■ **Cut back** any dead, damaged and diseased stems or branches to healthy live wood.
■ **Remove branches that cross** others or grow towards the centre of the plant; bushes, in particular, should be kept open-centred in a goblet shape.
■ **Further shorten sideshoots** on cordons, espaliers and fans, which were summer-pruned to four or five leaves, to just one or two buds. For gooseberries and red and white currants, prune back the tips of main branches at the same time.
■ **Thin out congested spurs** on mature trained forms of apple and pear; shorten long spurs by half, and remove any that are growing closely together.
■ **Encourage sideshoots** and spurs to develop on young espalier and fan-trained fruit by shortening

Prune apple trees in winter, when they are completely dormant.

main branches by half. Always cut to a downward-facing bud.
■ **On overgrown or neglected trees**, cut out one or two main branches to admit more light and air and to encourage new growth.
■ **Dispose of any diseased material**, preferably by burning it. Wash hands and clean tools before moving to another plant.

TREATING PESTS AND DISEASES
There are other measures you can take to protect plants from pests and counteract several serious diseases. Blackcurrant gall mites overwinter inside buds; check bushes in winter and early spring for the characteristic fat, rounded buds. Pick off any you find, and burn or otherwise destroy them. To combat apple and pear canker, prevalent on wet soils, remove diseased and mummified fruits, prune back affected shoots and cut out lesions on main stems. Use fungicidal wound paint on the cuts.

Painting or spraying fruit trees with a tar-oil winter wash is a traditional way of killing overwintering pest larvae and eggs, disease spores, moss and lichen. The wash is very toxic so before application protect the area under the trees with plastic sheeting or thick layers of newspaper. Bear in mind that it may kill beneficial insects as well as fruit pests.

Pruning gooseberries

1 Shorten to two buds all sideshoots that were summer-pruned to five leaves, and completely remove any thin or spindly shoots.

2 Cut out one or two of the old dark branches where young shoots are growing and tie these in, as replacements.

3 Remove surplus shoots growing from the base, clear weeds and mulch with straw, bracken or well-rotted manure.

Pear scab

- **PLANTS AT RISK** Pears.
- **RECOGNITION** Leaves and fruit form dark brown velvety blotches, and the fruit becomes distorted. The symptoms are caused by a fungus disease that overwinters on fallen leaves and also on the tree, so young shoots in spring may be blistered and cracked.
- **DANGER PERIOD** Growing season, but most severe after a wet May.
- **TREATMENT** See APPLE SCAB (p. 97).

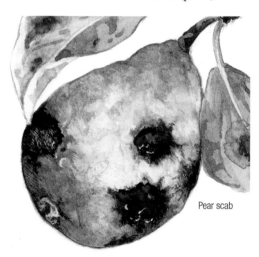

Pear scab

Potassium deficiency

▶ **SEE** page 62

Raspberry beetles

- **PLANTS AT RISK** Raspberries, loganberries and blackberries.
- **RECOGNITION** The tiny larvae of the raspberry beetle feed on the ripening fruit.
- **DANGER PERIOD** June to August.
- **TREATMENT** Spray thoroughly with derris as soon as flowering ceases.
Organic advice In autumn lightly fork over the soil around raspberries. Remove any mulch and replace it with fresh material in spring.

Raspberry beetle larvae

Raspberry virus

▶ **SEE** page 63

Russeting

- **PLANTS AT RISK** Apples, pears and other fruits.
- **RECOGNITION** Roughening of the skin – a natural characteristic, but sometimes a symptom of disease, such as powdery mildew, frost damage or chemical damage.
- **DANGER PERIOD** Any time.
- **TREATMENT** Control mildew, use chemicals carefully and ensure good cultivation. See FROST DAMAGE (p. 74) and POWDERY MILDEW (p. 62).

Russeting

Scald

▶ **SEE** page 109

Squirrels

▶ **SEE** page 76

Stony pit virus

■ **PLANTS AT RISK** Old pear trees.

■ **RECOGNITION** Fruit pitted and deformed, with patches of dead, stony cells in the flesh making it inedible. The virus appears first on one branch and then over the years spreads throughout the whole tree until all the fruit is affected.

■ **DANGER PERIOD** Any time.

■ **TREATMENT** Cut down and burn diseased trees.

Strawberry beetle

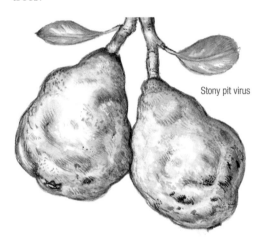

Stony pit virus

Strawberry beetles

■ **PLANTS AT RISK** Strawberries.

■ **RECOGNITION** Glossy-black ground beetles, 2cm long, eat pieces of the fruit as it ripens, causing damage similar to that made by birds. The pest may be found under the plants.

■ **DANGER PERIOD** Early summer.

■ **TREATMENT** Remove any debris and garden litter, remove weeds and maintain good hygiene. Spray with bifenthrin or derris, but not when plants are in bloom

Organic advice In winter, remove debris and litter.

Choosing resistant varieties

NO GARDEN IS ENTIRELY FREE OF DISEASE, but some varieties of plant are more susceptible than others, notably many roses, annuals, vegetables and fruit. Having to spray plants against diseases in order to keep them healthy is not only costly and time-consuming, but environmentally unfriendly too. Where possible, avoid encountering problems in the first place by choosing varieties that have good natural resistance to disease. Plant breeders have developed many new healthy cultivars in recent years. Particularly when choosing susceptible plants such as roses, check an up-to-date catalogue to seek out resistant varieties, or read descriptions on the labels to see whether disease resistance is mentioned before you buy.

Whatever the type of plant, bear in mind that healthy, well-grown specimens are much better able to resist attack than plants that are weak and under stress through lack of nutrients or water, or through being grown in the wrong conditions. Hence the importance of thorough soil preparation, choosing appropriate plants for the designated place in your garden, and providing good care for them as they become established in the first few months, or years in the case of woody plants.

Strawberry virus

Strawberry virus diseases

■ **PLANTS AT RISK** Strawberries.

■ **RECOGNITION** Discoloration of leaves due to virus diseases ARABIS MOSAIC VIRUS (p. 51) and CRINKLE (p. 54), which are spread by aphids.

■ **DANGER PERIOD** Any time.

■ **TREATMENT** Control the aphids (see p. 34). Destroy all diseased plants and replenish the stock with certified virus-free plants.

Wasps

■ **PLANTS AT RISK** The ripening fruits of apples, pears and plums.

■ **RECOGNITION** The wasps enlarge the damage already caused to fruit by birds.

■ **DANGER PERIOD** Late summer to early autumn.

■ **TREATMENT** Trace the nest and spray an anti-wasp insecticide at the entrance at dusk, when most of the wasps will be inside. **Organic advice** Wasps are excellent natural predators, controlling caterpillars and other pests. They should be killed only if it's absolutely necessary.

Wasp

Weevils

▶ **SEE** page 38

Anthracnose of beans

■ **PLANTS AT RISK** Dwarf and climbing french beans and sometimes runner beans.
■ **RECOGNITION** Sunken, dark brown patches occur on pods, and brown spots appear on the leaves. Caused by fungus.
■ **DANGER PERIOD** Growing season, particularly during cool wet weather.
■ **TREATMENT** Destroy all diseased plants and grow fresh plants from new seeds in a new position. Do not save seed from plants that may have the disease. If a serious outbreak reappears, spray before flowering with mancozeb.

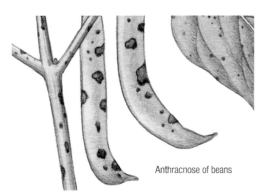

Anthracnose of beans

Bean seed flies

■ **PLANTS AT RISK** Beans, peas, sweetcorn and other vegetable crops.
■ **RECOGNITION** Seedlings fail to emerge. Small maggots eat germinating seeds and the stems of young seedlings.
■ **DANGER PERIOD** Seed-sowing time and as seedlings emerge.

■ **TREATMENT** Dust along the drill with derris when the seedlings emerge. **Organic advice** Raise plants indoors or cover newly sown seed with horticultural fleece or polythene to encourage quick growth.

Bean seed flies

Blight

■ **PLANTS AT RISK** Tomatoes and potatoes.
■ **RECOGNITION** In damp conditions, leaves develop yellow-brown blotches and a white furry coating underneath. They turn brown and rot. A dry brown rot on tomato fruits.
■ **DANGER PERIOD** May to August.
■ **TREATMENT** Spray every 10–14 days with mancozeb. Spray potatoes before the leaves touch other plants, and tomatoes when the first fruits set. **Organic advice** Grow resistant potatoes. As symptoms spread, cut off all foliage, and wait three weeks before harvesting.

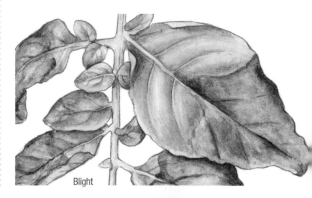

Blight

Blossom end rot

■ **PLANTS AT RISK** Tomatoes, peppers and aubergines.

■ **RECOGNITION** Circular brown or black patch at the blossom end of the fruit, which gradually enlarges to penetrate the flesh. Caused by a calcium shortage and dehydration.

■ **DANGER PERIOD** As fruit develops.

■ **TREATMENT** Prevent the soil from drying out and maintain even growth. This disease is a particular problem with plants that are grown in pots or growing bags, where they may need to be watered several times a day.

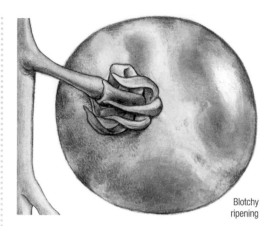

Blotchy ripening

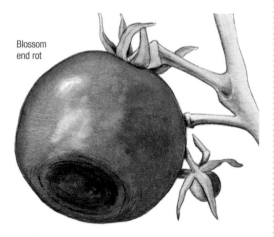

Blossom end rot

Blotchy ripening

▶ **SEE ALSO GREENBACK (p. 107)**

■ **PLANTS AT RISK** Tomatoes.

■ **RECOGNITION** Hard green or yellow patches on fruit, often on lower trusses.

■ **DANGER PERIOD** Growing season.

■ **TREATMENT** The problem may occur if plants have poor soil, are growing too fast under poor light, or it they are overwatered or too dry. A high temperature can also be a factor. Take appropriate action. A potash feed may help. Grow resistant varieties such as 'Alicante', 'Eurocross', 'Grenadier' and 'Shirley'.

Cabbage whiteflies

■ **PLANTS AT RISK** Cabbages, brussels sprouts and other brassicas.

■ **RECOGNITION** Leaves may be sticky and covered with sooty mould. Clouds of small white flies rise up when plants are disturbed.

■ **DANGER PERIOD** Growing season.

■ **TREATMENT** Spray the underside of leaves regularly with bifenthrin, fatty acids or pyrethrum. Do not spray within a week of harvest. Remove and burn infected plants after harvest. **Organic advice** Pick off leaves infested with young whitefly scales. Spray with insecticidal soap, on a cold day when whiteflies are less active.

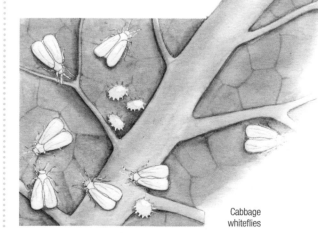

Cabbage whiteflies

Celery heart rot

■ **PLANTS AT RISK** Celery.

■ **RECOGNITION** When celery is lifted the centres have a wet, slimy brown rot, often extending up the stalk. Bacteria are believed to enter through wounds caused by slugs, or severe frost injury.

■ **DANGER PERIOD** Growing season.

■ **TREATMENT** Ensure good cultivation on well-drained soil and use a balanced fertiliser. Choose a fresh site for new celery to avoid a build-up of bacteria in the soil. Control slugs and give winter protection in cold districts. Carry out crop rotation.

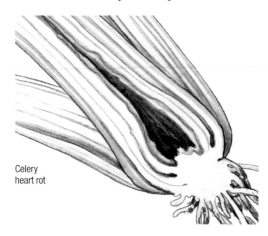

Celery
heart rot

Chocolate spot

■ **PLANTS AT RISK** Broad beans.

■ **RECOGNITION** Beans, foliage and sometimes the stems become covered with chocolate-coloured spots caused by a fungal disease. The whole plant may die.

Chocolate
spot

■ **DANGER PERIOD** June or July, or after severe winter frosts. Overwintered crops are most at risk in a wet spring.

■ **TREATMENT** Spray with Bordeaux mixture or mancozeb as foliage appears. Encourage strong growth by liming the soil if necessary, using a potash fertiliser, ensuring good drainage and sowing seeds thinly. **Organic advice** Don't sow autumn or early spring crops on wet soils. Don't sow or plant too close together.

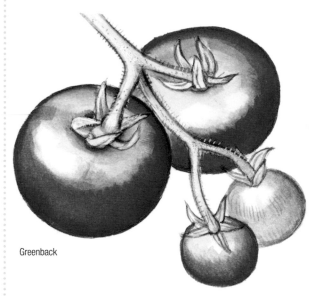

Greenback

Greenback

▶ **SEE ALSO BLOTCHY RIPENING (p. 106)**

■ **PLANTS AT RISK** Tomatoes.

■ **RECOGNITION** As fruit starts to ripen, a ring (partial or complete) of hard leathery tissue is left around the stalk end. This area does not turn red – see PHYSIOLOGICAL DISORDERS (p. 61).

■ **DANGER PERIOD** As fruit develops.

■ **TREATMENT** The major cause is thought to be exposure to heat, so even fruits shaded from the sun in a greenhouse can succumb. In hot weather, ventilate the greenhouse and provide shade. Maintain even growth by ensuring that the soil does not dry out. Alternatively, grow resistant varieties.

Weedkiller damage

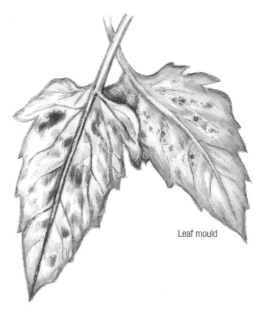

Leaf mould

Hormone weed-killer damage

■ **PLANTS AT RISK** All types, but tomatoes and vines are particularly susceptible.
■ **RECOGNITION** Leaves become narrow, often yellow, fan-shaped and frequently cupped.
■ **DANGER PERIOD** Growing season.
■ **TREATMENT** Plants usually recover without treatment. To avoid future damage, do not use weedkiller spray on a windy or hot day, or use weedkiller equipment for any other jobs. Mowings from a freshly treated lawn can be put on a compost heap, provided they are not spread on the garden for at least six months and are well decomposed.

Leaf mould

■ **PLANTS AT RISK** Tomatoes that are being grown under glass. Rare on outdoor plants.
■ **RECOGNITION** Purple-brown mould on the underside of leaves, with yellow blotches on the upper side. The plant's growth is checked and fruit develops poorly.
■ **DANGER PERIOD** Summer. Sometimes as early as April, but not usually until June.
■ **TREATMENT** Improve cultural conditions and maintain a maximum temperature in the greenhouse of 21°C with good ventilation. Do not let high temperature

coincide with high humidity. Remove and destroy lower foliage if it shows signs of attack. Check the disease by spraying with mancozeb or copper fungicide. Grow resistant varieties.

Pea and bean weevils

■ **PLANTS AT RISK** Young garden peas and beans; older plants are unaffected.
■ **RECOGNITION** Leaf edges are scalloped where eaten by the small beetle-like insects and their larvae.

Pea and bean weevil

■ **DANGER PERIOD** March to July.
■ **TREATMENT** Apply derris dust or spray as soon as the symptoms appear. **Organic advice** Spray with derris dust. Grow plants under horticultural fleece.

Pea moths

■ **PLANTS AT RISK** Garden peas.
■ **RECOGNITION** Moths lay eggs in the flowers and small maggot-like caterpillars feed on the peas inside the ripening pods.
■ **DANGER PERIOD** May to August.
■ **TREATMENT** Spray when flowers first open, and again two weeks later, with derris or permethrin. Alternatively, grow early-maturing varieties of peas.
Organic advice Sow peas early (February) or late (May) so that they are not in flower during June or July.

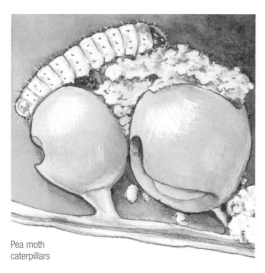
Pea moth
caterpillars

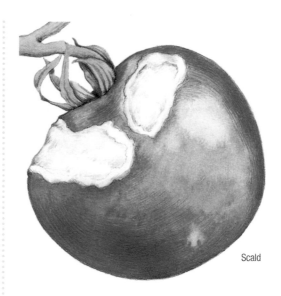
Scald

Scald

■ **PLANTS AT RISK** Plums, apples and soft fruit – particularly tomatoes and grapes under glass.
■ **RECOGNITION** Plums have red sunken patches; apples have discoloured patches on the skin encircled by a halo; tomatoes have creamy white wrinkled patches; grapes have discoloured patches.
■ **DANGER PERIOD** Long hot summers.
■ **TREATMENT** Remove affected fruits – even if the flesh beneath is undamaged – before GREY MOULD (p. 56) develops. Damp down the greenhouse early in the day so that any moisture on the fruit dries out before the sun is too strong. Shade greenhouse crops during bright periods.

Potassium deficiency
▶ **SEE** page 62

Stem rot
▶ **SEE** page 76

Tobacco mosaic virus

■ **PLANTS AT RISK** Primarily tomatoes, but also some cacti, orchids and herbaceous species.

■ **RECOGNITION** Various strains produce varied symptoms: dark and light green mottling on the leaves; bright yellow spots and patches on leaves and fruit stalks; and fruits blotched with brown.

■ **DANGER PERIOD** Growing season.

■ **TREATMENT** None. Burn all diseased plants and their roots, including debris in the soil. Eradicate greenhouse pests. Sow only tomato seeds that are certified to be free of virus infection.

Tobacco mosaic virus

Tomato spotted wilt virus

Tomato spotted wilt virus

■ **PLANTS AT RISK** Tomatoes under glass, houseplants and herbaceous species outdoors.

■ **RECOGNITION** A mosaic or mottling of the foliage, which may be distorted, due to a virus transmitted by thrips.

■ **DANGER PERIOD** Growing season.

■ **TREATMENT** None. Burn all infected plants and control thrips (see p. 84).

Cabbage root flies
▶ **SEE** page 86

Chrysanthemum stool miners
▶ **SEE** page 68

Leatherjackets
▶ **SEE** page 91

Root aphids and root mealy bugs

■ **PLANTS AT RISK** Cacti and succulents, primulas and other pot plants, lettuces and some outdoor ornamentals.

■ **RECOGNITION** White, wax-covered aphids or mealy bugs infest the roots of plants. Growth is checked, and the plants may turn yellow and wilt.

■ **DANGER PERIOD** Outdoors: late summer and autumn. Under glass: any time.

■ **TREATMENT** Water the non-edible plant roots with imidacloprid or thiacloprid. **Organic advice** Grow resistant varieties of lettuce. Mealy bugs are worse in dry conditions, so keep the soil or compost moist.

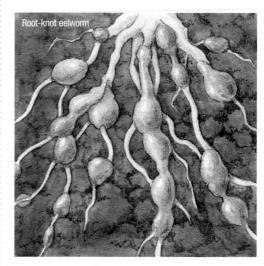
Root-knot eelworm

Root aphids and root mealy bugs

Root-knot eelworms

■ **PLANTS AT RISK** Cucumbers and tomatoes under glass, as well as other greenhouse and pot plants. Also some outdoor plants.

■ **RECOGNITION** Roots develop lumpy galls where eelworms are present which check and affect growth.

■ **DANGER PERIOD** Any time.

■ **TREATMENT** No chemical cure. Burn infested plants and wash contaminated pots in garden disinfectant.

Basal rot
of lilies

Basal rot of lilies
(or Fusarium rot)

- **PLANTS AT RISK** Lilies.
- **RECOGNITION** Growth either fails to appear above ground or is yellow and weak. Bulbs quickly rot and disintegrate.
- **DANGER PERIOD** Any time.
- **TREATMENT** Examine bulbs carefully before planting. Burn affected bulbs. Do not plant new lily bulbs in the same bed for several years.

Basal rot

- **PLANTS AT RISK** Crocuses, narcissi and onions.
- **RECOGNITION** Bulbs usually do not grow in spring, and the roots start to rot. If a corm or bulb is sliced downwards, it shows dark strands spreading from the chocolate-brown rot at the base, up through the inner scales.
- **DANGER PERIOD** Any time, including storage of narcissi bulbs.

- **TREATMENT** Destroy rotten corms and bulbs. **Organic advice** Grow resistant varieties of the bulbs.

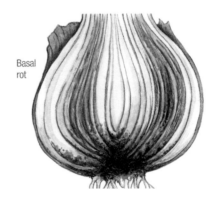

Basal
rot

Blue mould

- **PLANTS AT RISK** Stored vegetables, corms and bulbs.
- **RECOGNITION** Blue or blue-green spores of fungus, which are encouraged by a moist atmosphere, appear on plant material.
- **DANGER PERIOD** Any time.
- **TREATMENT** Ensure dry cool conditions and that all plant material is healthy when stored.

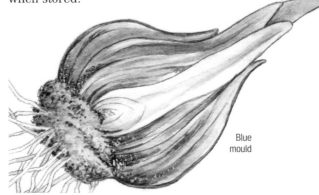

Blue
mould

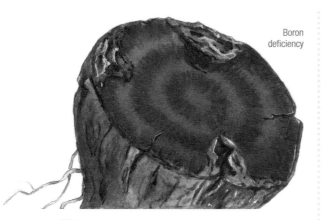

Boron
deficiency

Boron deficiency

■ **PLANTS AT RISK** Beetroots, swedes and celery.

■ **RECOGNITION** Edible roots turn brown inside; celery stalks develop brown cracks.

■ **DANGER PERIOD** May to August.

■ **TREATMENT** Mix 30g of borax with 10 litres of water and spread it over 10m² of soil.

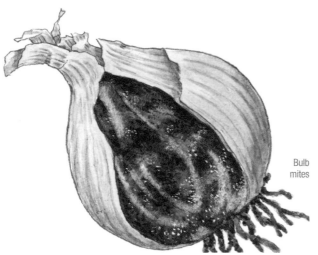

Bulb
mites

Bulb mites

■ **PLANTS AT RISK** Bulbs of narcissi, hyacinths, tulips, lilies, gladioli, dahlias and freesias.

■ **RECOGNITION** Small, pearl-white mites invade damaged tissue of the bulbs.

■ **DANGER PERIOD** Depends on the time of the damage.

■ **TREATMENT** Take care when lifting bulbs to avoid damaging them. Prevent damage by pests or disease.

Bulb scale mites

▶ **SEE** page 43

Carrot flies

■ **PLANTS AT RISK** Carrots, parsnips, parsley and celery.

■ **RECOGNITION** Reddening of foliage and sometimes stunted growth. Roots riddled with tunnels made by fly maggots.

■ **DANGER PERIOD** June to October.

■ **TREATMENT** In areas where the pest is prevalent, sow thinly in May. **Organic advice** Grow carrots under horticultural fleece or in pots or tubs at least 45cm above ground level so that adults cannot reach them. Early or late sowings will avoid the worst periods of attack. If possible, grow the crop in a windy spot. Intercrop with onions or dill – grow four rows of onions or dill for every row of carrots. Some varieties of carrot are partially resistant.

Carrot fly
maggots

Common scab

■ **PLANTS AT RISK** Potatoes, beetroots, radishes, swedes and turnips.

■ **RECOGNITION** Ragged-edged scabs develop on potato tubers due to organisms related to bacteria. The scabs do not render the potatoes inedible, but they should be removed before the potatoes are cooked.

■ **DANGER PERIOD** Growing season.

■ **TREATMENT** Burn peelings or severely infected tubers. Avoid liming the soil before sowing or planting; instead, add humus – such as green manure – and maintain even growth by keeping the soil moist and by mulching. Grow resistant varieties, such as 'Arran Pilot', 'King Edward' and 'Pentland Crown'. Put a layer of grass mowings in the bottom of the trench when planting.

Common scab

Core rot

■ **PLANTS AT RISK** Freesia and gladiolus corms.

■ **RECOGNITION** Rotting spreads from the centre outwards, as the corm becomes spongy and gradually turns brown or black.

■ **DANGER PERIOD** During storage.

■ **TREATMENT** Burn infected corms. After lifting corms, dust them with sulphur and store them in a dry atmosphere at 7–10°C.

Core rot

Cracking

■ **PLANTS AT RISK** All root vegetables, especially carrots and parsnips. Also apples, pears, plums, gages and tomatoes.

■ **RECOGNITION** Roots split lengthways; no symptoms appear above soil level. Fruit skins split to expose inner flesh.

■ **DANGER PERIOD** Growing season – dry periods followed by heavy rain.

■ **TREATMENT** Avoid irregular growth by mulching heavily and maintaining even watering so that the soil doesn't dry out.

Cracking

Dry rot

■ **PLANTS AT RISK** Potatoes and some other tubers; also crocuses, freesias and gladioli.

■ **RECOGNITION** Exposed growth turns yellow-brown and then collapses as leaf sheaths rot. Tubers (or corms) develop small dark lesions which enlarge and merge as the tubers shrivel and die. The cause is a soil-borne fungus.

■ **DANGER PERIOD** From January onwards and during storage.

■ **TREATMENT** Destroy all infected tubers or corms as soon as the first symptoms appear. Before storing or replanting healthy ones, dust ornamental corms and tubers with sulphur. Store them in a dry, cool, frostproof place. Plant tubers or corms in a fresh site each year.

Grey
bulb rot

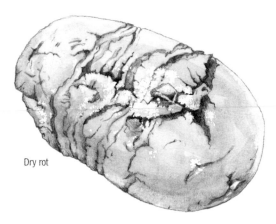

Dry rot

Gladiolus scab
▶ **SEE** page 56

Grey bulb rot

■ **PLANTS AT RISK** Tulips, hyacinths and other bulbous plants.

■ **RECOGNITION** Young shoots do not appear in spring. A dry grey rot at the neck of the bulbs gradually spreads to develop clusters of black fungi, and then the bulbs disintegrate.

■ **DANGER PERIOD** Soon after planting.

■ **TREATMENT** Remove and burn all debris from infected plants; replace surrounding soil with sterilised compost. Soak bulbs and corms in mancozeb before planting. Preferably avoid planting bulbs in infected areas for several years.

Ink disease
▶ **SEE** page 91

Hard rot

■ **PLANTS AT RISK** Chiefly gladiolus corms, but also the corms of crocus and freesia.

■ **RECOGNITION** Sharply defined tiny brown spots on leaves which develop infectious fungi in diseased tissue; black-brown sunken spots on corms. More common in the south of Britain than the north.

■ **DANGER PERIOD** Fungus spreads fast in wet weather in poor soils. Symptoms may appear in storage.

■ **TREATMENT** Destroy all diseased bulbs and corms, and burn the foliage at the end of the season as the disease may overwinter on dead plant debris. Dip healthy corms in a solution of mancozeb before storing.

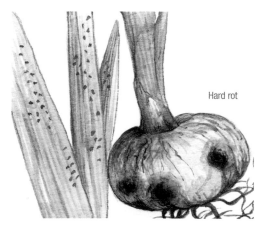

Hard rot

Internal
rust spot

Internal rust spot

■ **PLANTS AT RISK** Potatoes.

■ **RECOGNITION** Brown spots and blotches in the flesh of potato tubers, possibly due to dry soil.

■ **DANGER PERIOD** Growing season.

■ **TREATMENT** Incorporate plenty of well-rotted manure or compost into the soil before planting, and water regularly.

Mice and voles

■ **PLANTS AT RISK** Bulbs and corms, including crocus, lily, narcissus and tulip, in open ground or in storage. Peas and beans at sowing time.

■ **RECOGNITION** Damage and droppings.

Mice or vole
damage

■ **DANGER PERIOD** Any time between autumn and spring, especially with bulbs recently planted where the soil is still soft.

■ **TREATMENT** Before planting, dust bulbs and corms with a proprietary animal repellent. In storerooms lay bait containing coumatetralyl or difenacoum, but cover to avoid endangering birds and pets.

Organic advice Use rodent traps.

Millipedes

■ **PLANTS AT RISK** Root vegetables, bulbs, tubers and corms.

■ **RECOGNITION** Grey-black worm-like insects, similar to centipedes but slower in movement and with more legs, feed inside the rootstock of plants.

■ **DANGER PERIOD** Late summer to autumn.

■ **TREATMENT** Improve garden hygiene and dig deeply. Protect seeds, bulbs and corms with bendiocarb dust.

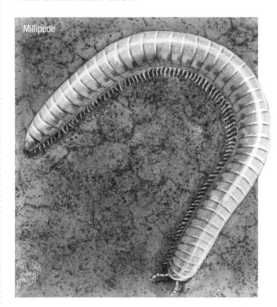

Millipede

Narcissus flies

■ **PLANTS AT RISK** Amaryllis, narcissi and snowdrops.

■ **RECOGNITION** Softening and rotting of bulbs as grubs burrow into them, particularly in dry soil. Affected bulbs fail to flower, and just produce narrow grass-like leaves.

■ **DANGER PERIOD** Late spring to early summer. Bulbs planted in sunny sites are most vulnerable.

■ **TREATMENT** Lift and burn affected bulbs and any grubs still in the soil. Apply derris dust to the soil before or soon after planting. **Organic advice** As foliage dies down, holes are left in the soil which allow narcissus flies to lay eggs near the bulbs. Rake soil into the holes. Alternatively, cover bulbs with horticultural fleece as the leaves die.

Neck rot

Narcissus fly larvae

Neck rot

■ **PLANTS AT RISK** Onions.

■ **RECOGNITION** A grey velvety mould develops near the neck of stored onions causing them to rot rapidly.

■ **DANGER PERIOD** Growing season, but symptoms do not appear until storage.

■ **TREATMENT** Destroy diseased onions as they appear. Store only those that are well ripened and hard in a dry, airy place. Use a three or four-year rotation system. Do not overfeed with nitrogen-rich manures or fertilisers as large bulbs are more prone to neck rot than smaller ones.

Parsnip canker

■ **PLANTS AT RISK** Parsnips.

■ **RECOGNITION** Reddish brown cankers may form on the shoulder of a root, with the leaves developing small black spots with green halos. The disease is caused by fungus.

■ **DANGER PERIOD** June onwards.

■ **TREATMENT** Avoid disease by sowing seeds early in deep loamy soil, enriched with balanced fertiliser, and lime if needed. Improve drainage and rotate crops. Control carrot fly (see p. 113). If trouble persists, grow resistant varieties. **Organic advice** Small roots, produced by growing parsnips at a spacing of 7.5–10cm, are less prone to canker. Earthing up may help.

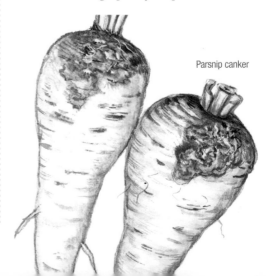
Parsnip canker

Potato-cyst
eelworms

Potato-cyst eelworms
(Golden nematode)

■ **PLANTS AT RISK** Potatoes and tomatoes.

■ **RECOGNITION** Plants grow poorly. If
infection is severe, they may wilt and die.
Roots reveal tiny white, yellow or brown
cysts that can only be seen through a
magnifying glass.

■ **DANGER PERIOD** May to August.

■ **TREATMENT** Dig up and burn affected
plants and avoid planting potatoes or
tomatoes on the same site for five years.
Organic advice Build up the organic level
in the soil by applying leaf-mould, compost
or well-rotted manure. Grow early potatoes
that may crop before eelworm levels build
up. Choose varieties that are resistant to
eelworm.

Powdery scab

■ **PLANTS AT RISK** Potatoes.

■ **RECOGNITION** Raised round scabs that
subsequently burst to release a powdery
mass of spores. Tubers may be deformed
by the soil-inhabiting fungus and have an
earthy taste.

■ **DANGER PERIOD** In heavy soils, during
cool damp weather throughout the growing
season.

■ **TREATMENT** Destroy diseased tubers.
Avoid planting potatoes on the same site
for several years. Improve drainage and
grow resistant varieties.

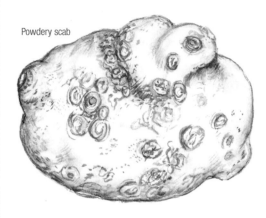

Powdery scab

Rhizome rot
▶ **SEE** page 93

Ring rot

■ **PLANTS AT RISK** Potatoes.

■ **RECOGNITION** Soft, cheese-like rot
encircling inner flesh. In severe cases,
skins may crack.

■ **DANGER PERIOD** Growing season

■ **TREATMENT** Notifiable.

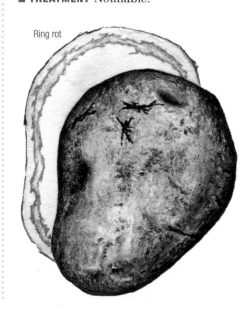

Ring rot

Soft rot

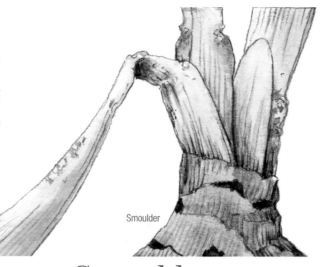

Smoulder

Smoulder

■ **PLANTS AT RISK** Narcissi.

■ **RECOGNITION** Leaves develop grey velvety mould and rot; bulbs decay – the result of a fungal disease. More severe in a cold, wet season.

■ **DANGER PERIOD** Foliage vulnerable in the spring; bulbs during storage.

■ **TREATMENT** Destroy infected bulbs and store remainder in a cool, dry place. Remove and burn affected plants during the growing season as soon as symptoms are seen. Spray the rest with mancozeb at 10-day intervals.

Smuts

▶ **SEE** page 48

Soft rot

■ **PLANTS AT RISK** Root vegetables and stored roots.

■ **RECOGNITION** A bacterial disease which reduces plants to a soft, slimy and foul-smelling rot. The disease enters the plant through damaged tissues.

■ **DANGER PERIOD** Any time.

■ **TREATMENT** Destroy infected plants and vegetables. Improve your cultural methods and storage conditions.

Squirrels

▶ **SEE** page 76

Stem and bulb eelworms

■ **PLANTS AT RISK** Daffodils, tulips, phlox, hyacinths, onions and many other plants.

■ **RECOGNITION** Leaves become swollen and distorted; bulbs crack and rot; plants die.

■ **DANGER PERIOD** Dormant bulbs in late summer and autumn; growing plants usually in spring.

■ **TREATMENT** Destroy affected plants and avoid replanting in the same soil for at least three years. Maintain good hygiene in the garden.

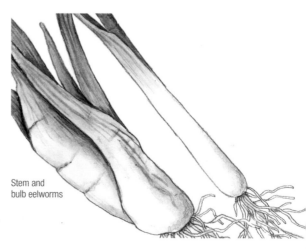

Stem and bulb eelworms

Swift moths

■ **PLANTS AT RISK** Carrots, parsnips and various herbaceous perennials.

■ **RECOGNITION** The caterpillars of the swift moth are dirty white with brown heads. They live in the soil and feed on roots, tubers, corms and rhizomes, and may eat up into the stems.

■ **DANGER PERIOD** During the growing season.

■ **TREATMENT** No chemical treatment
Organic advice Usually a problem only on newly cleared land. Thorough winter digging will expose the caterpillars and kill them. Good and consistent weed control and regular cultivation will help to reduce the risk.

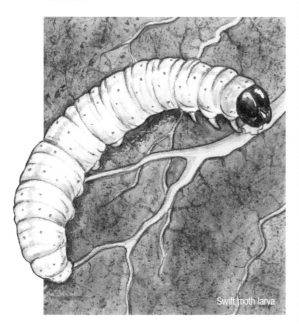

Swift moth larva

Tulip bulb aphids

■ **PLANTS AT RISK** Crocuses, gladioli, irises and tulips.

■ **RECOGNITION** Colonies of dark green aphids on stored bulbs and corms.

■ **DANGER PERIOD** During storage.

■ **TREATMENT** Before storing, dip bulbs in a

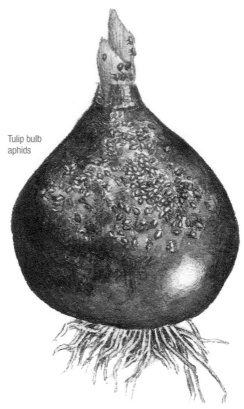

Tulip bulb aphids

spray-strength solution of bifenthrin and allow to dry out. Remove mild infestations of aphids by hand.

Tulip fire

▶ **SEE** page 84

Wart disease

■ **PLANTS AT RISK** Potatoes.

■ **RECOGNITION** Developing tubers produce large warty outgrowths, and subsequently disintegrate, due to a soil-borne fungus.

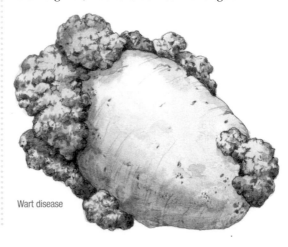

Wart disease

■ **DANGER PERIOD** Growing season.

■ **TREATMENT** Notifiable. All new potato varieties are immune to the disease, but some older ones are not. All diseased potatoes must be destroyed.

Violet root rot
▶ **SEE** page 95

Weevils
▶ **SEE** page 38

White rot

■ **PLANTS AT RISK** Mainly salad onions, and occasionally leeks, shallots and garlic.

■ **RECOGNITION** Bulbs develop a white fungus at their base and rot.

■ **DANGER PERIOD** Growing season.

■ **TREATMENT** Burn the affected plants. Grow onions on a new site each year – once the disease has struck, the soil remains contaminated for up to 20 years. No chemical treatment is available to amateur gardeners.

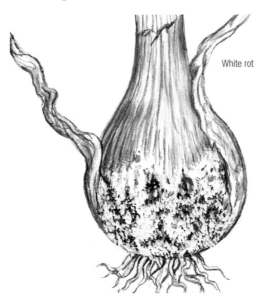

White rot

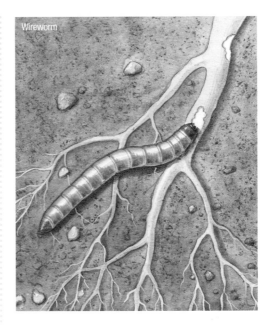

Wireworm

Wireworms

■ **PLANTS AT RISK** Potatoes, lettuces, tomatoes and dendranthemas (chrysanthemums).

■ **RECOGNITION** Yellow-brown larvae live in the soil to feed on the roots of potato tubers, and also on the stems of other plants.

■ **DANGER PERIOD** The early years of newly cultivated land.

■ **TREATMENT** No chemical treatment. Reduce wireworm colonies by frequent cultivation of the soil.

Fairy ring

Fairy rings

■ **PLANTS AT RISK** Lawns.

■ **RECOGNITION** Inner and outer rings of lush, dark green grass appear in the lawn, with brown or dead turf between them. In summer and autumn – especially during wet weather – small brown-capped toadstools flourish between the rings.

■ **DANGER PERIOD** Any time.

■ **TREATMENT** No effective chemical control is available. **Organic advice** Improving the growing conditions of your lawn – by spiking to improve drainage, for example – might help, together with applications of sulphate of iron. Feeding the lawn heavily with a nitrogen-rich fertiliser disguises the effect of the ring.

Leatherjackets

▶ **SEE** page 91

Lichens

■ **PLANTS AT RISK** Grass, trees and shrubs.

■ **RECOGNITION** Grass: overlapping leaf structures which grow horizontally in the turf – a deep green-black colour when moist, and grey-green or brown when dry. Trees and shrubs: thin crusts of grey or orange tissue, or leafy-looking plants, growing on the bark.

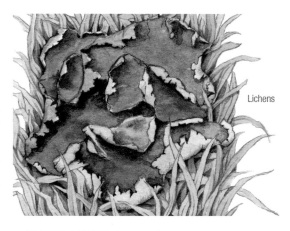
Lichens

■ **DANGER PERIOD** At any time.

■ **TREATMENT** Grass: rake out the growths. Moss-killer products may reduce the problem. Drain if possible and spike the surface. Brush in lime-free sand, top-dress and feed. Trees and shrubs: lichen is harmless and should be left. Encourage vigorous growth of the host plant by feeding in spring and summer.

Red thread (Corticium)

■ **PLANTS AT RISK** Grass.

■ **RECOGNITION** Dead patches of turf with outbreaks of red fungus.

■ **DANGER PERIOD** Autumn, after rain.

■ **TREATMENT** Spray with fungicide according to manufacturer's instructions.

Red thread

Aerate the soil and apply a nitrogenous fertiliser in spring.

Snow mould

- **PLANTS AT RISK** Grass.
- **RECOGNITION** Most obvious after a snow thaw or during moist weather; a white, cotton-like growth of fungus covers patches of turf, which turn brown and die.
- **DANGER PERIOD** Any time, but most severe in October.
- **TREATMENT** Treat as for RED THREAD (p. 122). After August, apply an autumn lawn fertiliser only, which is low in nitrogen but high in phosphate and potash.

Snow mould

Moss control

To control moss successfully, you must tackle the underlying causes such as shade, compacted soil and poor drainage. Moss can also build up in wet winters on otherwise healthy lawns and, if left untreated, it will smother and eventually kill the grass. The use of a lawn sand, that is a combined moss and weedkiller with added fertiliser, is an efficient solution. As this product feeds the grass, it will recover quickly and grow over the gaps left by the dead moss.

Choose a day when there is heavy dew so that the chemicals stick to the moss and weed leaves. The moss and other weeds will turn black within five to seven days, but wait until the moss turns brown, which indicates that it is completely dead, before raking. Any patches that return to green should be treated again. Collect and dispose of the dead moss, but not in the compost bin, because the chemical residues will taint its contents.

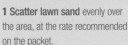

1 Scatter lawn sand evenly over the area, at the rate recommended on the packet.
2 Use a fan-shaped wire rake to remove all dead brown moss and debris from the lawn.
3 Spiking the lawn in late spring will aerate the soil and help to reduce moss, which thrives in damp lawns.

THERE IS MORE TO GROWING A LAWN THAN SIMPLY LAYING THE TURF AND MOWING IT ONCE A WEEK. A SMOOTH AND HEALTHY-LOOKING LAWN REQUIRES SOME TIME SPENT CONTROLLING WEEDS AND FEEDING AS WELL AS TRIMMING AND MOWING. YOUR EFFORT WILL BE REWARDED.

MOWING THE LAWN

Lawns do not need much care in the winter, but once the soil temperature rises above 5–7°C, the grass starts growing and you need to mow. The exact timing will vary from year to year, depending on the weather. Make the first cuts with the mower blades set high. As the rate of growth increases, the lawn will need more frequent mowing, and the height of cut can gradually be lowered. Aim to reduce the grass by a third of its height at each cutting; do not cut it too short, as this weakens the grass and exposes bare soil where moss and weeds can easily establish.

If you prefer not to use chemicals, prise out the weeds individually with an old kitchen knife or a tool called a 'grubber'.

SPRING FEEDING

Spring is also the time to start feeding your lawn. Use a spring or summer fertiliser that is high in nitrogen and phosphates to promote growth and root development. If you use a powdered or granular formulation, rather than liquid feed, wait for a day when the grass is dry and the soil is moist, so that the fertiliser settles on the soil. If there is no rain within 48 hours, water the lawn lightly to wash in the fertiliser.

CONTROLLING WEEDS AND MOSS

The most effective way to control lawn weeds is with a combination of chemical and cultivation practices, involving weedkiller use in conjunction with frequent mowing.

- **Use a fan-shaped rake** to lift low-spreading weeds, such as speedwell and yarrow, before mowing, so that the blades slice off their topgrowth.
- **Kill tap-rooted weeds** like dandelion and dock by digging down and cutting them off about 8cm below soil level.
- **Control spreading weeds** by applying a spray or granular lawn weedkiller, which will kill broad-leaved weeds without harming the grass.
- **Brush off worm casts** as these contain weed seeds brought up from below ground, which will germinate when exposed to the light. Always brush the lawn before mowing, using a stiff brush to scatter the casts. This helps to reduce wear and tear on mower blades caused by the gritty nature of the casts and prevents them being smeared over the lawn by the mower, smothering the grass beneath and causing a patchy appearance.

Dealing with bare patches

Bare patches may be due to weeds smothering the grass, or an area being covered with an object. Spills of concentrated fertiliser or other chemicals may 'burn' the grass and kill it off in patches. Fortunately, these areas can be re-seeded to restore the lawn to its original condition.

1 First go over the area with a spring-tined rake. Rake vigorously to drag out all the old dead grass, including pieces of dead root, and to score the soil surface. Use a garden fork to break up the surface. Jab the tines into the soil to a depth of 2–3cm.

2 Rake the soil to a depth of 1–2cm to level it and to create a fine seedbed.

3 Sow grass seed evenly over the area, at a rate of about 30g per m², and lightly rake it in.

4 Cover the area with sacking until the seeds germinate; water in dry weather.

Acknowledgments

All images in this book are copyright of The Reader's Digest Association Limited. The majority previously appeared in *New Encyclopedia of Garden Plants & Flowers* with the exception of the following which appeared in *1001 Hints & Tips for the Garden* 41B, 46BR, 68B, 83T, 118B & 120T, and *Short Cuts to Great Gardens* 11 & 13.

Wood cut designs by John Woodcock

Photography
The position of photographs on each page is indicated as follows: **T** top; **B** bottom; **L** left; **R** right; **C** centre.
Front Jacket www.istockphoto.com/Tomasz Resiak; **TL & spine** iStockphoto.com/www.pmsicom.net; **1, 2-3** Digital Vision; **4 T** © RD; **5 B** FLPA/Nigel Cattlin; **6-7** © RD; **8** © RD/Mark Winwood; **9, 10** © RD; **12, 15, 17 C** © RD/Mark Winwood; **18** © RD/Sarah Cuttle; **19 T** © RD/Mark Winwood; **19 B** © RD/Sarah Cuttle; **20, 21 B** © RD/Mark Winwood; **22** © RD/Sarah Cuttle; **23** © RD/Mark Winwood; **24** © RD/Debbie Patterson; **25 T** © RD/Sarah Cuttle; **25 B** © RD/Debbie Patterson; **26 & 27 T** © RD/Mark Winwood; **27 B, 28 TL & TR** © RD/Sarah Cuttle; **28 B, 29** © RD/Mark Winwood; **30-31** FLPA/Nigel Cattlin; **32** iStockphoto.com/Marion Habel; **35** © RD/Mark Winwood; **59 BC, CL** © RD/Sarah Cuttle; **59 CLB & T & B** © RD/Mark Winwood; **67** © RD/Mark Winwood; **70 B** iStockphoto.com/Debbie Martin; **97, 101, 123, 124** © RD/Mark Winwood

Reader's Digest Slugs, Pests & Diseases is based on material in *Reader's Digest New Encyclopedia of Garden Plants & Flowers, 1001 Hints & Tips for the Garden* and *All Seasons Guide to Gardening*, all published by The Reader's Digest Association Limited, London.

First Edition Copyright © 2007
The Reader's Digest Association Limited,
11 Westferry Circus,
Canary Wharf,
London E14 4HE

www.readersdigest.co.uk

Editor Alison Candlin
Art Editor Austin Taylor
Assistant Editor Celia Coyne
Editorial Consultant Geoff Stebbings
Proofreader Ron Pankhurst
Indexer Marie Lorimer

Reader's Digest General Books
Editorial Director Julian Browne
Art Director Nick Clark
Managing Editor Alastair Holmes
Head of Book Development Sarah Bloxham
Picture Resource Manager Sarah Stewart-Richardson
Pre-press Account Manager Sandra Fuller
Senior Production Controller Deborah Trott
Product Production Manager Claudette Bramble

Origination Colour Systems Limited, London
Printed and bound in China by CT Printing

ISBN: 978 0 276 44204 9
BOOK CODE: 400-616 UP0000-1
ORACLE CODE: 250010676H.00.24